Alkaloid Chemistry

Alkaloid Chemistry

MANFRED HESSE

Institute of Organic Chemistry,
University of Zurich

Translated from the German Edition by
I. Ralph C. Bick, University of Tasmania

A WILEY-INTERSCIENCE PUBLICATION
JOHN WILEY & SONS   New York   ●   Chichester   ●   Brisbane   ●   Toronto

Originally published in 1978 as *Alkaloidchemie*.
© 1978 Georg Thieme Verlag.

English translation Copyright © 1981 by John Wiley & Sons, Inc.

Library of Congress Cataloging in Publication Data:

Hesse, Manfred, 1935-
  Alkaloid chemistry.

  Translation of Alkaloidchemie.
  "A Wiley-Interscience publication."
  Bibliography: p.
  Includes index.
  1. Alkaloids. I. Title.
QD421.H5413   547.7'2   80-22828
ISBN 0-471-07973-1

Printed in the United States of America

10 9 8 7 6 5 4 3 2 1

In memory of my illustrious teacher,
Professor Hans Schmid

Preface

Alkaloid chemistry constitutes at present a highly active field of research, with bases constantly being isolated and new results being produced, and for this reason it is hardly possible to write a comprehensive monograph on the subject: many volumes would be required for such a purpose. The present text seeks instead to give the student some concept of the aims and methods of research in the alkaloid field. In consequence, comparatively few data on individual alkaloids are included, since these are constantly being extended, updated, or rendered obsolete. This books provides instead an introduction to methods of continuing importance in alkaloid research, and is designed to stimulate the student's interest in this fascinating branch of natural products chemistry. The text is thus intended primarily for advanced students of organic chemistry as an introduction to the complex chemistry and spectroscopy of alkaloids.

I am most grateful to Dr. D. Ganzinger and to Mr. A. Guggisberg, both of the University of Zürich, for their help in the preparation of certain sections of this book. I also would like to express my thanks to the *Schweizerische National-fonds zur Förderung der wissenschaftlichen Forschung*, which supported this work.

MANFRED HESSE

Zurich, Switzerland
March 1981

Contents

Alkaloid Chemistry

Introduction

The first sections of this book deal with general aspects of alkaloid chemistry: the definition of an alkaloid (Section 2), problems of nomenclature (Section 3), and the problem of artifacts (Section 4). Section 5 provides a review of the more important types of skeleton found in naturally occurring alkaloids; at the same time this section gives an indication of the diversity of alkaloid structures. In the subsequent sections, methods of investigation that are of general utility are discussed by way of a study of specific examples.

For most structural types it must be reailized that each class of alkaloid, even each individual base to some extent, has its own separate chemistry from the point of view of structural determination, synthesis, and biogenesis. We cannot go into all the details of each structural type for lack of space, but since the methods of investigating these questions are the same for practically all alkaloids, methodology will occupy a central position in the treatment. This applies, for instance, in the discussions on alkaloid biogenesis (Section 6), chemotaxonomy (Section 7), structural elucidation by modern methods (Section 8), and degradative reactions (Section 9).

In comparison with the amount and diversity of the synthetic work on alkaloids described in the literature, the section on this subject (Section 11) is rather short; the four examples of synthesis discussed constitute only a limited selection from a field that has received a great deal of attention in recent years. In the section on dimeric alkaloids and bisalkaloids (Section 10) the main interest lies in the method of coupling of monomeric bases to produce complex ones.

A bibliography is provided at the end of the book, and in addition references to review articles and textbooks are given in Sections 5 and 10. Special references are provided in the remaining sections as required. For each alkaloid discussed in the text, the plant genus from which it was isolated is given. The names of plant families are quoted in accordance with those in reference W 14.

Concept and Definition of an Alkaloid

It is more difficult than might at first be supposed to define the term alkaloid. The work was coined in 1818 by Meissner and implies a compound similar to an alkali (from Arabic *qalay* = to roast and Greek εῖδος = appearance), referring to the basic properties of this class of substance.

The following definition is given in Meyer's *Konversations-Lexikon* of 1896: "Alkaloids (plant bases) occur characteristically in plants, and are frequently distinguished by their remarkable physiological activity. They contain carbon, hydrogen, and nitrogen, and in most cases oxygen as well; in many respects they resemble the alkalis (hence the name)." Definitions in modern dictionaries differ only in minor details from those of the older nontechnical literature.

In the scientific literature, difficulties in defining the term alkaloid were pointed out many years ago. In 1931, Trier wrote: "As a result of scientific progress, collective terms of this kind (i.e., alkaloids) will have to be abandoned in the end" [T 1]. The concept is nevertheless still in use today, in spite of problems such as the following: strychnine (2/1), isolated from *Strychnos* spp.

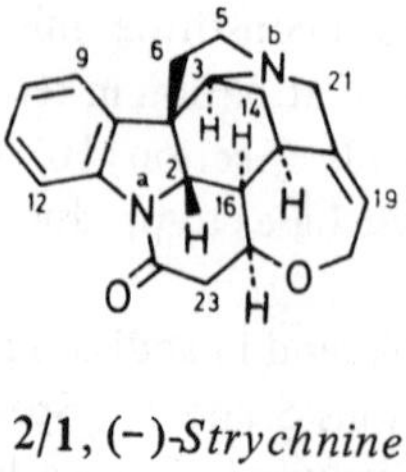

2/1, (−)-*Strychnine* 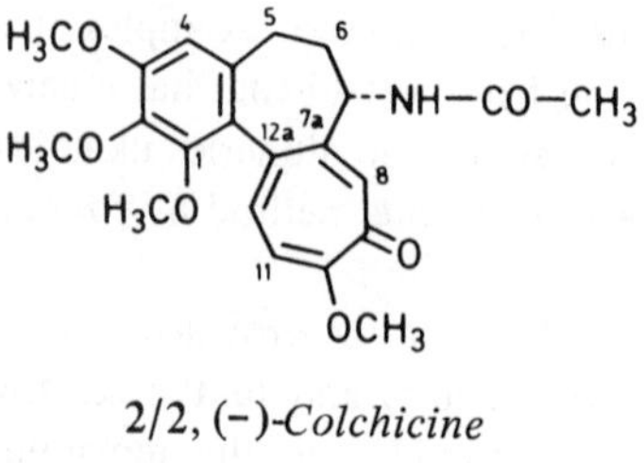   2/2, (−)-*Colchicine*

[e.g., *S. nux-vomica* (STR)] has a pK_a value of 8.26 and must be considered a strong plant base: it agrees with the definition cited above. On the other hand, colchicine (2/2) is an *N*-acetyl derivative and has almost neutral properties [from *Colchicum autumnale* (LIL)]. Various other types have feebly developed basic properties, such as the amide or lactam alkaloids [e.g., ricinine (2/3) from *Ricinus communis* (EUP)] and *N*-oxides [e.g., kopsinoline (2/4) from *Pleiocarpa*

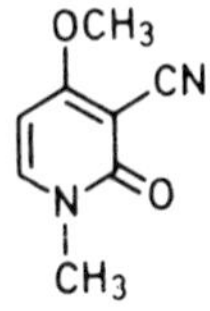

2/3, *Ricinine*

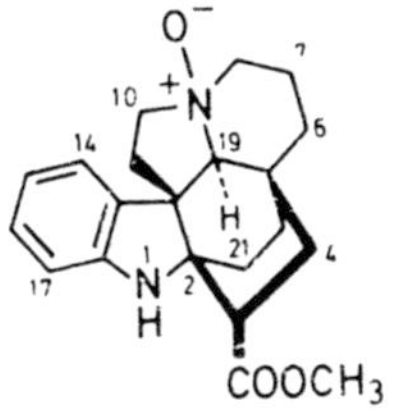

2/4, *Kopsinoline*

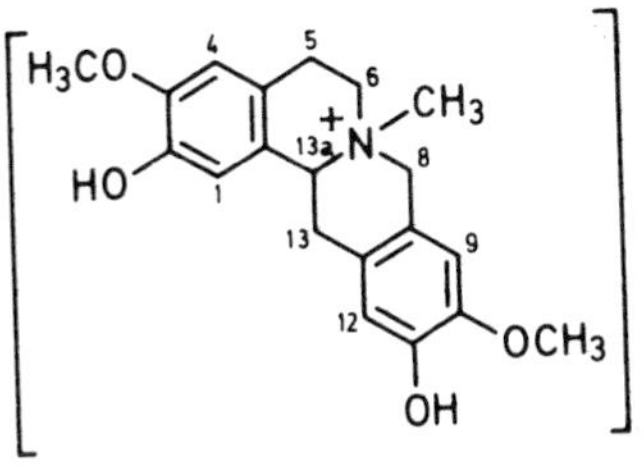

2/5, (±)-*Phellodendrine iodide*

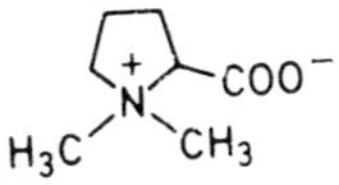

2/6, (±)-*Stachydrine*

mutica (APO)]; also quaternary nitrogen compounds [e.g., phellodendrine iodide (2/5) from *Phellodendron amurense* (RUT)], and betaines [e.g., stachydrine (2/6) from *Stachys sieboldi* (LAB)]. In contrast to their extensive occurrence in the plant kingdom, alkaloids have so far been found only rarely in animals: for example, samandarine (2/7), the main alkaloid of the poison gland in the skin of *Salamandra maculosa*.

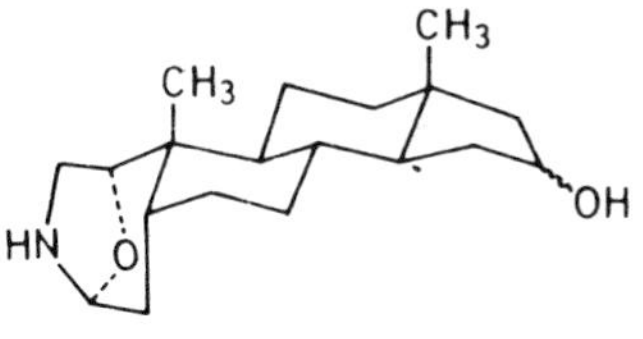

2/7, *Samandarine*

These examples suggest a different definition for alkaloids: simply nitrogen-containing compounds derived from either animals or plants. But again we would have to restrict the term, since amino acids, peptides, and nucleotides are not considered to be alkaloids; on the other hand, one speaks of peptide alkaloids, a group to which the ergot alkaloids belong. These are lysergic acid derivatives isolated from certain fungi, such as ergotamine (2/8), from *Claviceps purpurea* (CLA), which contains the amino acids L-α-hydroxyalanine, L-phenylalanine, and L-proline in the peptide portion. Apart from these types, simple derivatives of ammonia such as triethylamine and propylamine are not included among the true alkaloids, and for practical reasons the purines, pyrimidines, pyrazines, pterines, vitamins and their derivatives, as well as amino sugars and

2/8, *Ergotamine*

$$H_2N\text{-}(CH_2)_4\text{-}NH_2$$

2/9, *Putrescine*

$$H_2N\text{-}(CH_2)_4\text{-}NH\text{-}(CH_2)_3\text{-}NH_2$$

2/10, *Spermidine*

$$H_2N\text{-}(CH_2)_3\text{-}NH\text{-}(CH_2)_4\text{-}NH\text{-}(CH_2)_3\text{-}NH_2$$

2/11, *Spermine*

antibiotics are likewise excluded. On the other hand, derivatives of the simple aliphatic di-, tri-, and tetraamino compounds [e.g., putrescine (2/9), spermidine (2/10), and spermine (2/11)] are considered to be alkaloids, since a group of compounds, mostly macrocyclic with the above-mentioned amines as building blocks, have recently been discovered. Compounds 2/9, 2/10, and 2/11 belong to the group of so-called biogenic amines, which includes tryptamine (2/12) and other alkaloid building blocks, and which are likewise not considered to be alkaloids [C 12], [G 7].

2/12, *Tryptamine*

Finally, it will be observed that the oxidation number of the central nitrogen atom in nearly every case is −III, the same as that of the amines. But there are also *N*-oxides [e.g., kopsinoline (2/4)] and hydroxylamine derivatives [e.g., nareline (2/13), from *Alstonia scholaris* (APO)] with an oxidation number of −I, and—on biogenetic grounds—nitro derivatives where it reaches +III [e.g.,

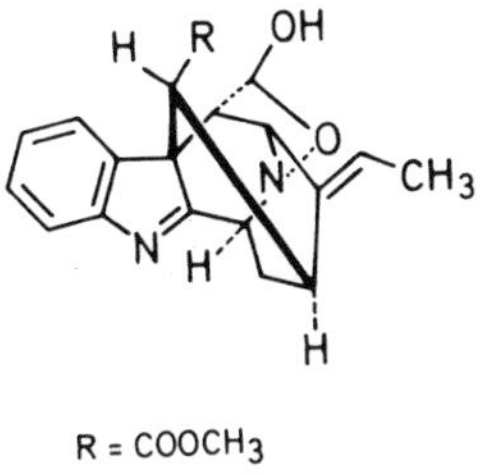

R = COOCH₃

2/13, *Nareline*

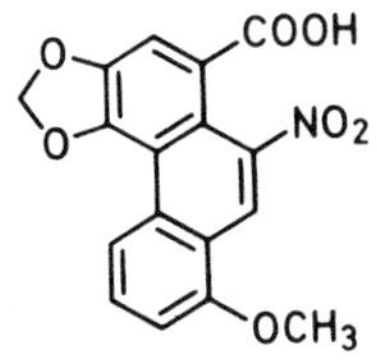

2/14, *Aristolochic acid I*

aristolochic acid I (2/14) from *Aristolochia* sp. (ARI)]. Clearly the division of naturally occurring nitrogen-containing substances into alkaloids and non-alkaloids is somewhat arbitrary, and the boundary line will be drawn in different positions by different authors according to their research interests.

Nomenclature

As in the case of other classes of natural products (e.g., flavonoids, terpenoids), no uniform system of nomenclature has so far been devised for alkaloids. In most cases the name of the alkaloid has been derived from the plant name (usually the systematic name). Thus papaverine was called after the *Papaver* species from which it was isolated. The names cocaine (from *Erythoxylum coca*) and atropine (from *Atropa belladonna*) arose in a similar way, but spegazzinine (from *Aspidosperma chakensis* Spegazzini) is named after the botanical authority. Frequently several alkaloids are obtained from the same plant, and the names devised for them will depend on the inspiration of the natural products chemist who isolated them. The usual ending is -ine in English and French and -in in German, but various other suffixes such as -idine, -anine, -aline, -inine, and so on, may be added to the same stem (e.g., the *Lobelia* alkaloids: lobeline, etc.). Alternatively, letters or numbers may be affixed to the name of the main alkaloid to designate a minor alkaloid. Naturally occurring derivatives of the main alkaloid are seldom clearly indicated as such in the names ascribed to them. This is partly due to the fact that the structural relationships were established only after the names had been allotted.

Some effort is at present being made to indicate the ring system in the stem of the alkaloid name and to designate newly discovered plant bases systematically as derivatives of this fundamental type. It remains to be seen how far this system will be adopted; it would be highly desirable for a computer-based program.

Artifacts

The term artifact is applied to a synthetic product that does not occur as such in nature. Artifacts that resemble alkaloids may be formed either from naturally occurring alkaloids or from natural products that do not contain nitrogen. They cannot be recognized as such from their structures alone, since the latter are often of a type that could occur in nature.

A wide variety of chemical reactions can lead to artifact formation. In working up plant material for the isolation of alkaloids, various extraction and concentration steps have to be carried out during which solvents, acids, and bases are brought into contact with the crude plant material. Furthermore, oxygen of the air is rarely excluded during the process. In view of the great diversity of alkaloid structures, it is not surprising that chemical reactions can take place between the plant material on the one hand and the reagents on the other. The following three examples are chosen by way of illustration from a large number of such reactions.

1 In addition to various alkaloids isolated from the apocynaceous plant *Hedranthera barteri*, the compounds 1,2-dehydrobeninine (**4/1**) and the base (**4/2**) were also obtained [A 4]. The compound **4/2** differs from **4/1**

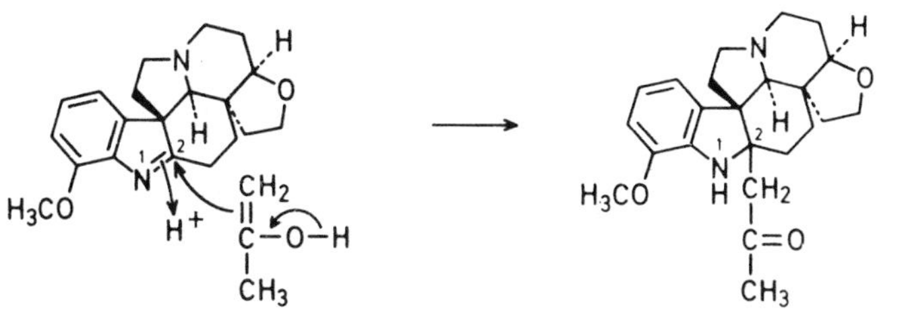

4/1, 1,2-*Dehydrobeninine*   4/2

7

in containing an additional mole of acetone, but if it is treated with 2*N* aqueous hydrochloric acid, 4/1 is formed. Furthermore, a model compound with a methoxyindolenine structure, which is present also in 4/1, forms the corresponding acetonyl derivative under conditions of acid catalysis in the presence of acetone. From these observations it appeared likely that the acetone used in the extraction and working up of the plant material could have reacted with 1,2-dehydrobeninine (4/1) to give 4/2, and that 4/2 could be an artifact. In consequence, an extraction of the plant material was carried out in the absence of acetone, and, as expected, the acetonyl compound 4/2 could no longer be isolated.

2 In many cases, aerial oxidation is responsible for the formation of artifacts. It is much more difficult in such cases, as compared to the first example, to determine whether or not an alkaloid that has been isolated is in its original form, as illustrated by the cases of vasicoline (4/3) and vascicolinone (4/4). These bases were obtained [J 6], along with others, from juvenile growth of the acanthaceous plant *Adhatoda vasica*. During the workup of the vasicoline-containing fraction, vasicolinone (4/4) was obtained. The same happened when pure vasicoline (4/3) was allowed to stand for a few hours in

4/3, *Vasicoline*   4/4, *Vasicolinone*

solution in the presence of air. The quantitites of the two bases isolated from the plant were about the same. By a carefully conducted workup of the plant material in the absence of air, it was possible to show that both alkaloid 4/3 and 4/4 were synthesized by the plant. On the other hand, it is likely that if *A. vasica* were worked up incorrectly, no 4/3 at all might be isolated, but only the oxidation product 4/4.

3 The base gentianine (4/5) was isolated from a number of plants of the families Gentianaceae and Loganiaceae. A biogenetically related substance, gentiopicroside (4/6), also occurs in plants of the Gentianaceae. Furthermore, swertiamarine (4/7), from plants of the genus *Swertia* (GEN), is closely related to 4/5, and both 4/6 and 4/7 contain a labile cyclic acetal grouping. On treatment with ammonia under mild conditions, both bases are converted into gentianine (4/5). This facile reaction suggests the possibility that gentianine is formed from 4/6 or 4/7 during the working up of the plant material

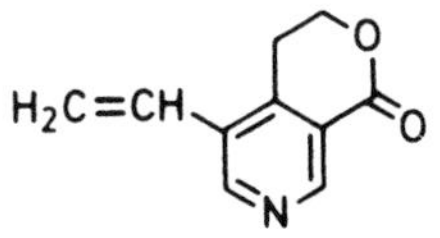

4/5, *Gentianine* 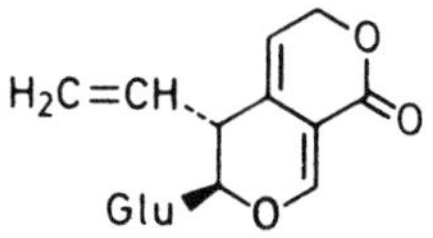   4/6, *Gentiopicroside* 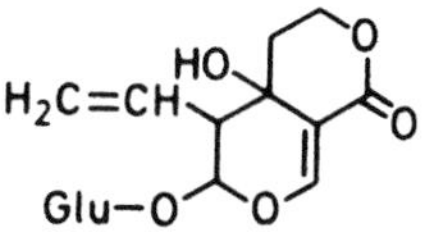   4/7, *Swertiamarine*

with ammonia. In consequence, a number of plants of the family Gentianaceae that had been considered up to that time to contain gentianine, were re-extracted using sodium carbonate instead of ammonia. This study revealed that only *Gentiana fetisowii* contains gentianine in large quantities; all the other plants investigated, contrary to the earlier studies, contained the base only in traces or not at all [F 4], [G 10]. It follows that most of the gentianine (4/5) that had been isolated previously was an artifact formed during the workup of the plant material with ammonia. It is not certain whether the plant contained gentianine originally, since even when the work-up is carried out without adding ammonia, sodium carbonate reacting with the plant material may set free ammonia, which could then be responsible for the formation of the gentianine.

Other examples of artifact formation which may be mentioned include transesterifications with ethanol, which is usually added to commercial grades of chloroform as a stabilizer; epimerization reactions in the presence of acids or bases; and quaternizations with chloroform.

It is doubtful whether artifact formation can be completely prevented, but one should in any case be aware of the possibility that one of the above-mentioned or similar reactions might occur during the extraction and workup of natural products—not only of alkaloids. For this reason it is advantageous, in carrying out an extraction, to set aside a sufficient quantity of the raw material to allow it to be worked up under different conditions if necessary after the structural determination work has been completed.

Classification of Alkaloids

Main Classes of Alkaloids

5.1	Heterocyclic Alkaloids
5.1.1	Pyrrolidine Alkaloids
5.1.2	Indole Alkaloids
5.1.3	Piperidine Alkaloids
5.1.4	Pyridine Alkaloids
5.1.5	Tropane and Related Bases
5.1.6	Histamine, Imidazole, and Guanidine Alkaloids
5.1.7	Isoquinoline Alkaloids
5.1.8	Quinoline Alkaloids
5.1.9	Acridine Alkaloids
5.1.10	Quinazoline Alkaloids
5.1.11	Izidine Alkaloids
5.2	Alkaloids with an Exocyclic Nitrogen
5.3	Putrescine, Spermidine, and Spermine Alkaloids
5.4	Peptide Alkaloids
5.5	Diterpene Alkaloids
5.6	Steroidal Alkaloids

Bibliography[1]

Reviews and collected works: [B 1], [B 15], [B 20], [C 6], [C 7], [C 11], [H 1], [H 17], [I 2], [M 1], [P 1], [R 1], [S 1], [T 1]

[1]Most of the references given are to review articles.

Biogenesis: [B 16], [L 1], [L 2], [L 7], [M 4], [S 7]

Mass spectra: [B 7], [B 8], [H 24], [R 9]

Chemotaxonomy: [H 2]

Spectra: [H 3], [Y 1]

Special reviews: [P 10], [P 11]

UV spectra: [S 33]

Pharmacology: [A 1]

Many alkaloids have been isolated from plants, animals, or native drugs (calabash curare, opium, etc.), or have been detected in these sources, and the large numbers involved make it necessary to subdivide them in some way. Roughly estimated there are over 5000 alkaloids of all structural types known at present. No class of naturally occurring organic substances shows such an enormous range of structures as the alkaloids; the structures of the steroids, flavonoids, and saccharides in comparison are confined to a relatively few fundamental types of skeleton. No uniform generally accepted principle exists whereby the alkaloids may be subdivided into classes without overlaps or contradictions; all the methods of classification that have so far been advanced involve intermediate cases for which compromises have to be made. The boundaries between different classes may be drawn according to the nature of the classification: for instance, the principle utilized may be based on biogenesis, structural relationships, botanical origin, or spectroscopic criteria (e.g., chromophore for UV spectroscopy, or ring skeleton for mass spectroscopy). Alkaloids may thus be subdivided into types with a distinct chromophore or fundamental skeleton, such as the indole, isoquinoline, or quinoline alkaloids. It may seem paradoxical at first sight that some of these types, such as isoquinoline, may include certain members that do not in fact have an isoquinoline residue, but nevertheless still belong to the group (e.g., the rhoeadine alkaloids). As a result of examples of this kind, one has to extend the fundamental "skeleton definition" and one has to say: alkaloids with an indole, isoquinoline, or quinoline nucleus, for example, and *biogenetically related bases*.

Besides this system of classification there is a method that depends on plant origin. This system has been applied to describe an alkaloid type of which the first example was isolated from a certain plant; examples are provided by the tobacco alkaloids and the amaryllidaceous alkaloids. However, representatives of these types are not confined to one plant family. Thus nicotine has been found not only in species of tobacco (Solanaceae), but also in a whole range of plants that have no relationship to tobacco. Another disadvantage of this type of classification is that the individual representatives can include a great variety of skeletal types (e.g., the amaryllidaceous alkaloids).

It is clear that the types of classification based on the fundamental skeleton or plant origin are mutually overlapping and do not permit a neat separation into classes. Another method is that based on a predominant bond type. Thus the peptide alkaloids are macrocyclic peptides with an additional amino component which gives them a basic character. Ergot alkaloids (German: *Mutterkorn-Alkaloide*) have lysergic acid as the basic component. On the other hand, this group of compound is considered to belong to the indole rather than to the peptide alkaloids, another interesting example of how boundary lines come to be drawn. The so-called biogenic amines, whose individual representatives are of very different skeletal types, form a further example. It is noteworthy that each new comprehensive work on alkaloids invariably has a chapter headed "Miscellaneous or Unclassified Alkaloids" as a result of the lack of a generally applicable principle of classification. This further emphasizes the problems involved in the concept and definition of an alkaloid. The structural demarcations between individual alkaloids on the one hand and peptides, amino acid derivatives, antibiotics, animal defence substances, steroids, and terpenes on the other constitute further problems.

In the survey of alkaloids that follows we shall be concentrating on the nitrogen atom and its immediate environment, and we shall attempt to classify them on the basis of their nitrogen-containing structural features. In cases where an alkaloid has more than one nitrogen atom, preference will be given to the most characteristic function. Exceptions to this principle will only be made when other structural elements are more typical, such as steroidal, terpene, spermidine, spermine, and peptide alkaloids. We can draw up the following classification scheme according to the location of the nitrogen atom in certain structural features:

1 Heterocyclic alkaloids.
2 Alkaloids with exocyclic nitrogen and aliphatic amines.
3 Putrescine, spermidine, and spermine alkaloids.
4 Peptide alkaloids.
5 Terpene and steroidal alkaloids.

When we classify the whole range of alkaloids according to this system, we find that they are divided up unequally: the great majority fall into the heterocyclic group, and the smallest group is formed by the putrescine, spermidine, and spermine bases.

5.1 HETEROCYCLIC ALKALOIDS

The representatives of this preponderant class of alkaloid have the nitrogen atom in a heterocyclic ring, and we shall begin with the five-membered ring.

Alkaloids with larger rings will be treated next, followed by compounds with two nitrogen atoms and by heterocyclic compounds containing a benzene ring fused on one side. Finally, bases in which the nitrogen is simultaneously a member of two ring systems will be discussed.

5.1.1 Pyrrolidine Alkaloids [H 5], [M 23], [M 24], [P 6]

In addition to the *N*-acetyl derivatives of pyrrolidine [e.g., 1-(3-methoxycin-namoyl)-pyrrolidine (5/1) from *Piper* sp. (PIP)], various alkylated pyrrolidines belong to this group: hygrine (5/2) from *Erythroxylum* sp. (ERY) serves as an example. Mesembrenol (5/3) from *Sceletium* sp. (AIZ) and dendrobine (5/4) from *Dendrobium* sp. (ORC) also contain nitrogen in a five-membered ring. Dendrobine and its derivatives are also considered to be sesquiterpene alkaloids from their carbon skeletons.

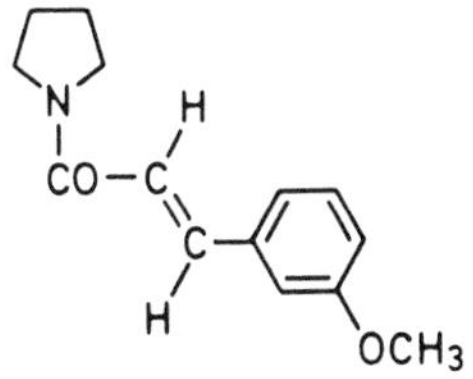

5/1, *1-(3-Methoxycinnamoyl)-pyrrolidine* 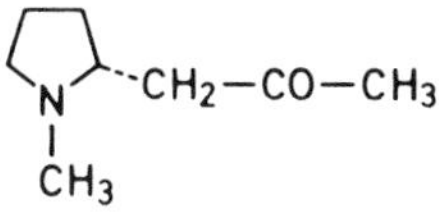   5/2, *(+)-Hygrine*

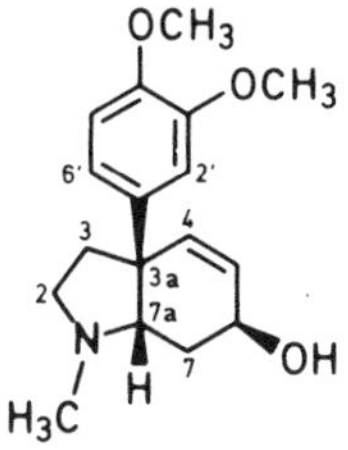

5/3, *Mesembrenol* 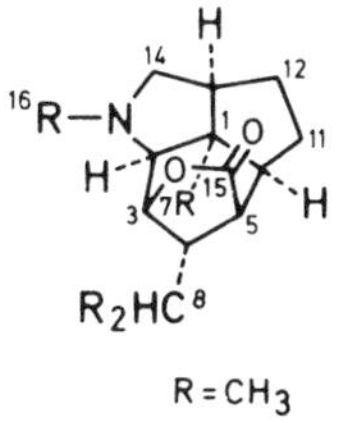   5/4, *(−)-Dendrobine*

5.1.2 Indole Alkaloids

Bibliography

General reviews: [B 17], [M 14], [N 1], [S 13], [S 34], [T 9]
Reviews of structures and lists of references: [H 14], [H 15]
Mass spectra: [H 16]
Synthesis: [K 7], [W 16], [W 17]

Biogenesis: [K 5], [S 39]

Special reviews:

Oxindole alkaloids: [B 35]; Alkaloids from *Alstonia*: [S 3], [S 4], [S 52];
Apocynaceae: [R 8]; *Aspidosperma*: [G 1], [G 2]; *Diplorrhyncus*: [G 1],
[G 2]; *Geissospermum*: [M 3]; *Gelsemium*: [S 12]; *Haplophyton*: [S 6];
Iboga: [T 13], [T 14]; *Kopsia*: [G 1], [G 2]; *Melodinus*: [G 1], [G 2];
Mitragyna: [S 16], [S 17], [S 51]; *Ochrosia*: [G 1], [G 2]; *Ourouparia*:
[S 16], [S 17]; *Picralima*: [S 21], [S 22], [S 52]; *Pleiocarpa*: [G 1], [G 2];
Pseudocinchona: [M 28]; *Vinca*: [T 10], [T 11]; *Voacanga*: [T 13],
[T 14]; *Yohimbe*: [M 28], [M 29].

There are about 1400 indole alkaloids known altogether, including compounds with the true indole chromophore (**5/5**) and those derived from it, such as indoline (dihydroindole, **5/6**), indolenine (**5/7**), hydroxyindolenine (**5/8**), α-methyleneindoline (**5/9**), pseudoindoxyl (**5/10**), and oxindole (**5/11**). Various derivatives of these nuclei **5/5**-**5/11** are included as well, such as those with a

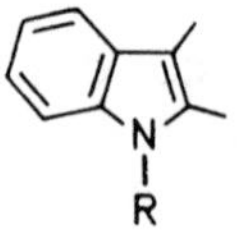

5/5, *Indole* 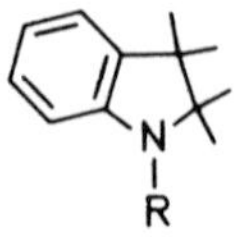   5/6, *Indoline or dihydroindole* 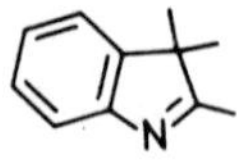   5/7, *Indolenine*

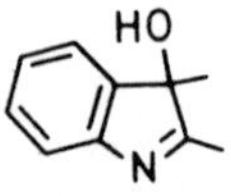

5/8, *Hydroxyindolenine* 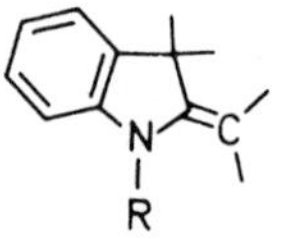   5/9, *α-Methyleneindoline* 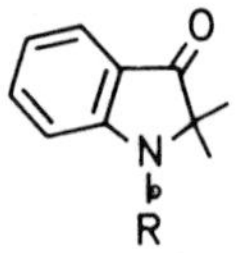   5/10, *Pseudoindoxyl*

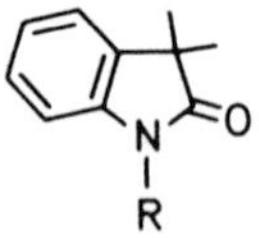

5/11, *Oxindole*

benzene or pyridine ring fused on [e.g., carbazole (**5/12**) and β- and γ-carboline (**5/13** and **5/14**)] and their substitution products. The indole alkaloids are further subdivided according to structural and biogenetic considerations.

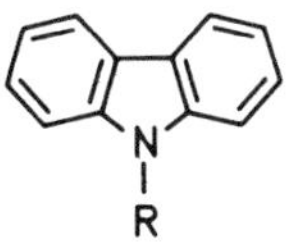

5/12, *Carbazole* 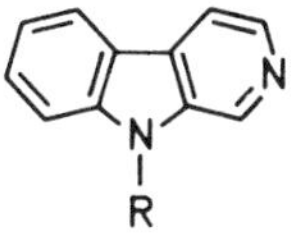   5/13, *β-Carboline* 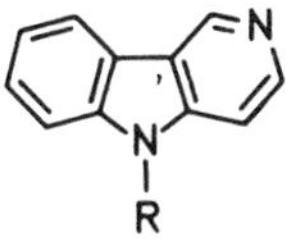   5/14, *γ-Carboline*

5.1.2.1 Simple Indole Bases [S 5], [S 14], [S 31]

The simplest natural representatives with an indole nucleus are the biogenic amines tryptamine (5/15) and serotonin (5/16). Gramine (5/17), which has a side chain shortened by one methylene group, also belongs to this type. A derivative of tryptamine, psilocybin (5/18, from *Psilocybe mexicana*), has hallucinogenic properties. Examples of other simple indole bases are apparicine (5/19, from various genera of the Apocynaceae) and vallesamine [5/20, from *Vallesia* sp. (APO)] .

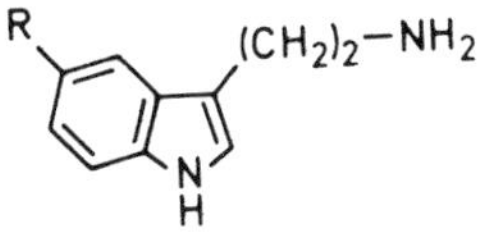

R = H   R = OH

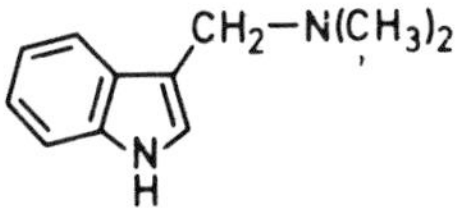

5/15, R = H   5/16, R = OH   5/17, *Gramine*
   Tryptamine   *Serotonin*

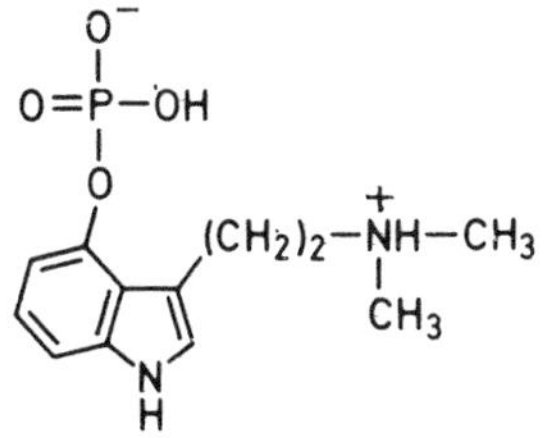

5/18, *Psilocybin* 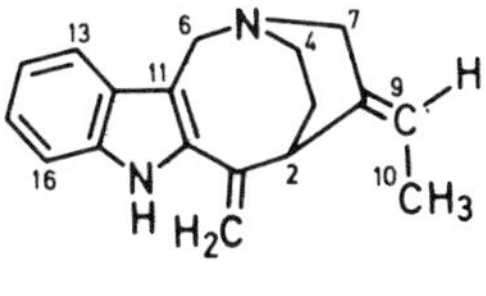   5/19, *Apparicine*

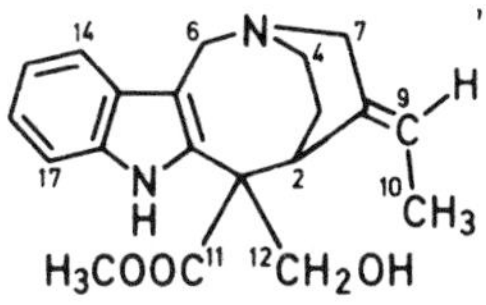

5/20, *Vallesamine*

5.1.2.2 Alkaloids with a Carbazole Nucleus

Most of the alkaloids in this section, which have a carbazole or β-carboline nucleus, are derived biogenetically from tryptamine (5/15).

5.1.2.2.1 Carbazole Alkaloids [K 13] A simple example of this type of alkaloid is glycozoline [5/21, from *Glycosmis* sp. (RUT)]. Other carbazole derivatives have been isolated from the genus *Murraya* (RUT), such as mahanimbine (5/22). A further group of compounds is derived from carbazole (5/12) by fusing on a pyridine ring. Olivacine [5/23, from various *Aspidosperma* spp. (APO)] is of this type.

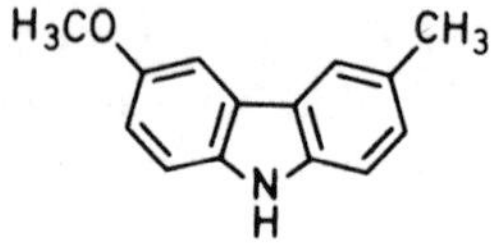

5/21, *Glycozoline* 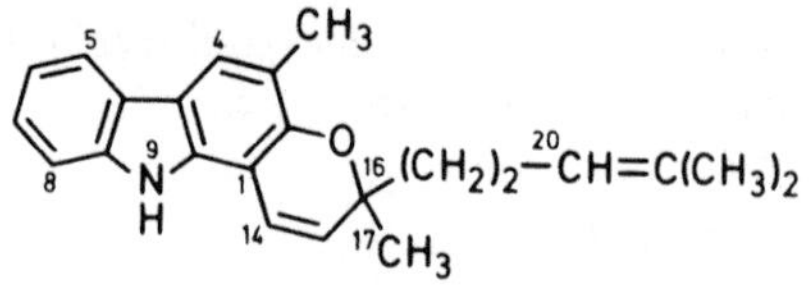   5/22, *Mahanimbine*

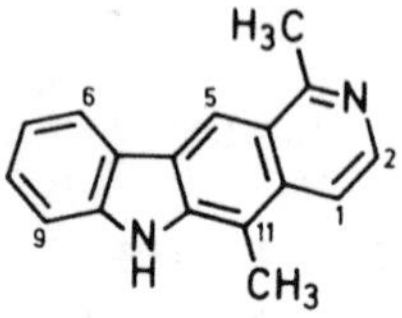

5/23, *Olivacine*

5.1.2.2.2 Aspidospermine and Biogenetically Related Bases (for plumerane alkaloids; see Section 7.2) About 250 alkaloids belonging to this subgroup have been isolated so far, all of which occur exclusively in the family Apocynaceae (for their chemotaxonomy; see Section 7.2). The following are examples of the various skeletal types:

15,16,17,20-Tetrahydrosecodine (5/24, from *Rhazya* sp.)

Quebrachamine (5/25, from *Amsonia, Aspidosperma, Gonioma, Melodinus, Pleiocarpa, Rhazya, Stemmadenia,* and *Vinca* spp.)

Aspidospermidine (5/26, from *Amsonia, Aspidosperma, Gonioma,* and *Rhazya* spp.)

Vincadifformine (5/27, from *Amsonia, Melodinus, Rhazya, Tabernaemontana, Vallesia,* and *Vinca* spp.)

Aspidoalbine (5/28, from *Aspidosperma* sp.)

Pleiocarpine (**5/29**, from *Hunteria* and *Pleiocarpa* spp.)

Tuboxenine (**5/30**, from *Pleiocarpa* sp.)

Kopsine (**5/31**, from *Kopsia* sp.)

Fruticosine (**5/32**, from *Kopsia* sp.)

Neblinine (**5/33**, from *Aspidosperma* sp.)

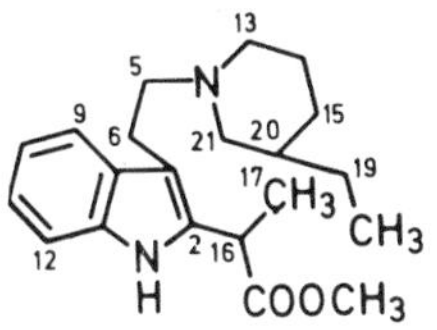

5/24, 15,16,17,20-*Tetrahydrosecodine* 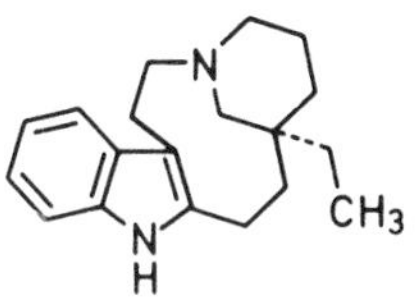   **5/25**, (+)-*Quebrachamine*

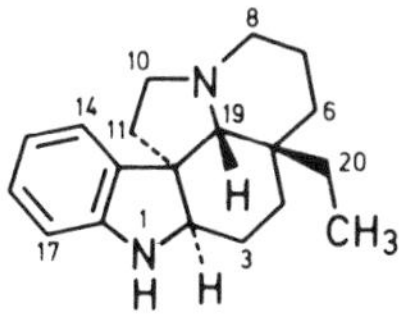

5/26, (+)-*Aspidospermidine* 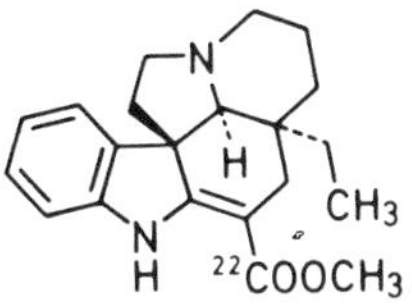   **5/27**, (−)-*Vincadifformine* 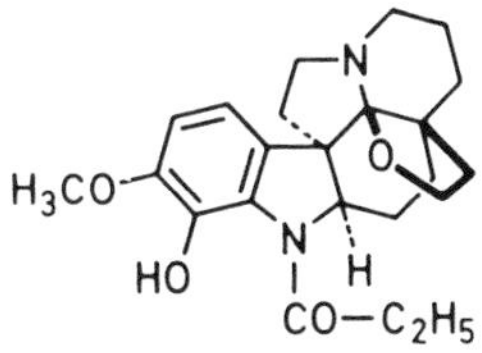   **5/28**, (+)-*Aspidoalbine*

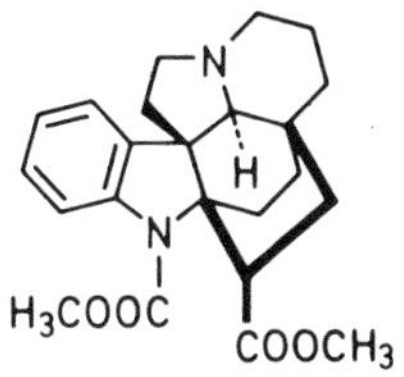

5/29, (−)-*Pleiocarpine* 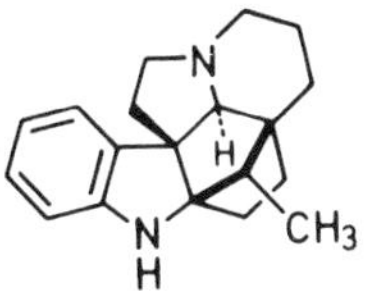   **5/30**, *Tuboxenine* 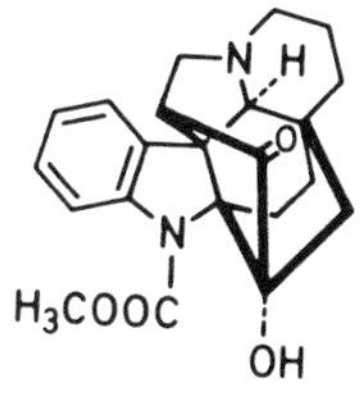   **5/31**, *Kopsine*

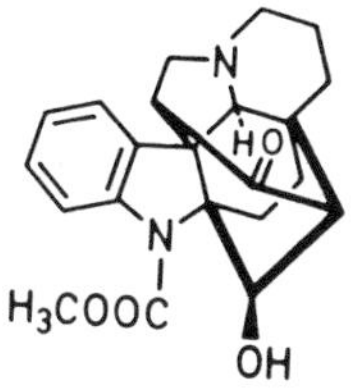

5/32, (−)-*Fruticosine* 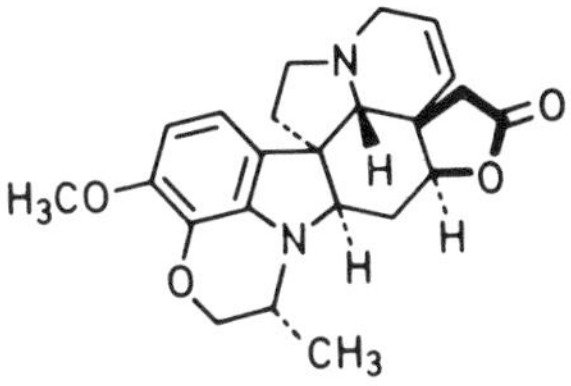   **5/33**, (−)-*Neblinine*

A whole range of analogous bases is derived from these fundamental types. With the exception of a few, which will be referred to later, they show no variation in the carbon-nitrogen skeleton, but they may have additional double bonds or oxygen functions, *N*-substitution, or ether rings [e.g., beninine (5/34) from *Hedranthera* sp.]. Vincatine (5/35, from *Vinca* sp.) is derived from alkaloids of the vincadifformine type by rupture of the C-2—C-3 bond. Aspidodispermine (5/36, from *Aspidosperma* sp.) differs from the aspidospermidine alkaloids in having no alkyl side chain, and pandoline [5/37, from *Pandaca* sp. (APO)] has this side chain in a different position (see Section 7.1).

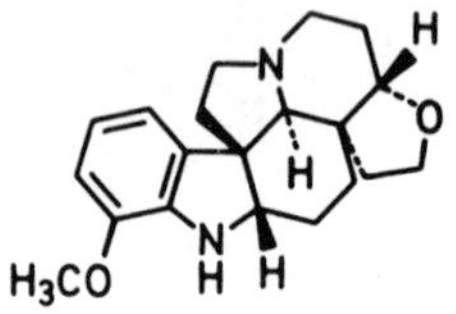

5/34, *Beninine*

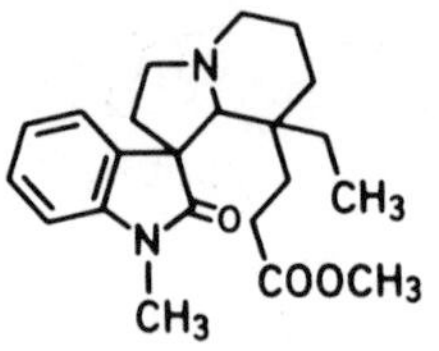

5/35, *Vincatine*

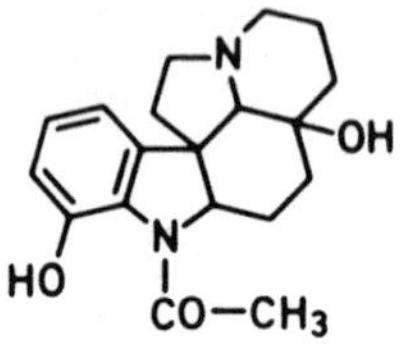

5/36, *Aspidodispermine*

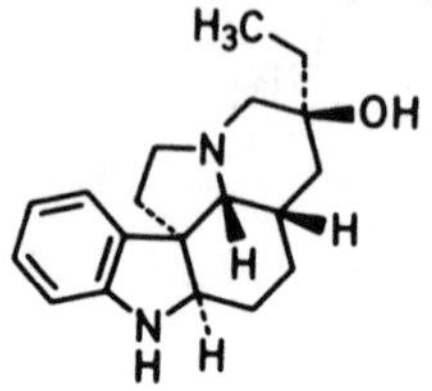

5/37, (+)-*Pandoline*

A large number of bisindole alkaloids (see Section 10) have a structure with one or two aspidospermine units.

5.1.2.2.3 Alkaloids of the Condylocarpine Type The most important representative of this group is condylocarpine [5/38, from *Diplorrhynchus* sp. (APO)]; most of the others differ from 5/38 in having no C-substituent at C-16.

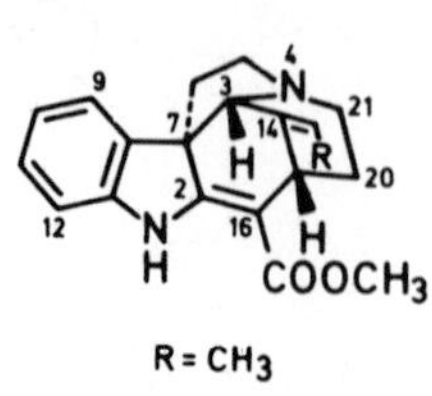

R = CH₃

5/38, (+)-*Condylocarpine*

H₃CO   R=CH₃

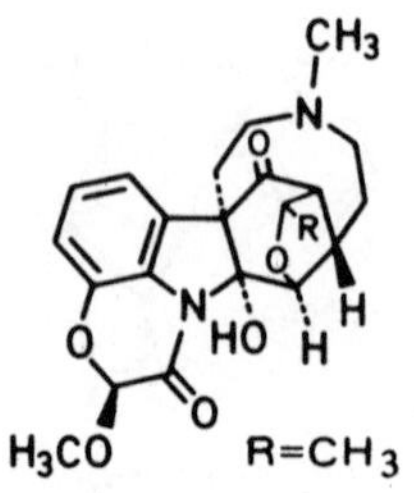

5/39, (+)-*Dichotine*

Precondylocarpine, on the other hand, has a hydroxymethyl group attached to this carbon atom in addition to the methoxycarbonyl group. Dichotine [5/39, from *Vallesia* sp. (APO)] has two additional rings.

5.1.2.2.4 Strychnos Alkaloids and Related Bases [B 10], [B 11], [B 12], [B 18], [H 11], [H 12], [H 13], [S 29]

A large number of indole alkaloids belong to this group; the most important skeletal types are represented by the following:

Stemmadenine [5/40, from *Diplorrhynchus, Melodinus, Stemmadenia,* and *Vallesia* spp. (APO)]

Tubifolidine [5/41, from *Pleiocarpa* sp. (APO)]

Akuammicine [5/42, from *Picralima* and *Vinca* sp. (APO)]

Wieland-Gumlich aldehyde [5/43, from *Strychnos* sp. (STR)] and strychnine (see 2/1; from *Strychnos* sp.), whose molecule has been enlarged by addition of a C_2 unit (strychnine is a potent nerve poison); also vomicine (5/44, from *Strychnos* sp.)

Tsilanine [5/45, from *Strychnos* sp. (STR)] contains an additional ring as compared to akuammicine.

The calabash curare alkaloids (see Section 10-3) are derived from the Wieland-Gumlich aldehyde (5/43), a degradation product of strychnine. The naturally occurring derivatives of these bases have no additional carbon atoms in their skeletons, but occur in reduced or oxidized forms, or with the nitrogen alkylated or acylated.

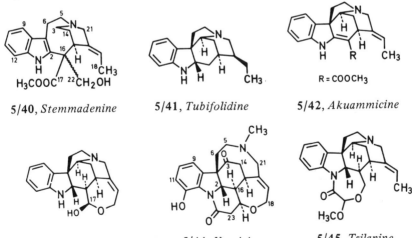

5/40, *Stemmadenine*   5/41, *Tubifolidine*   5/42, *Akuammicine*

5/43, *Wieland-Gumlich aldehyde*   5/44, *Vomicine*   5/45, *Tsilanine*

5.1.2.2.5 Uleine Alkaloids The condylocarpine type gives rise to a small group of bases in which the ethylene bridge between C-7 and the nitrogen N-4 (see 5/38) is lacking. An example is provided by uleine [5/46, from *Aspidosperma* sp. (APO)] .

R = ^{15}CH₃

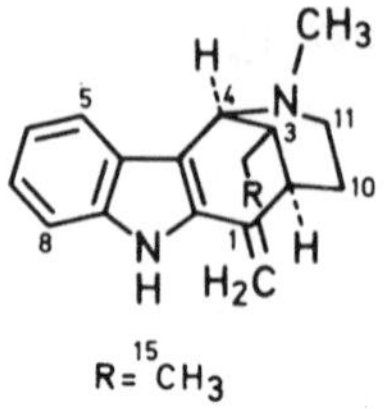

5/46, *Uleine*

5.1.2.3 Alkaloids with a β-Carboline Skeleton

5.1.2.3.1 Simple β-Carboline Derivatives [M 6] Low molecular weight derivatives of β-carboline are widely distributed in nature; thus harman (5/47) occurs in six plant families: *Arariba*, *Ophiorrhiza*, and *Sickingia* (RUB), *Caligonum* (POL), *Passiflora* (PAS), *Symplocos* (SYM), *Zygophyllum* (ZYG), and *Elaeagnus* (ELA). Many examples have additional carbon substituents on the nucleus, such as harman-3-carboxylic acid [from *Aspidosperma* sp. (APO)] . Brevicolline [5/48, from *Carvex* sp. (CYP)] contains a third nitrogen atom.

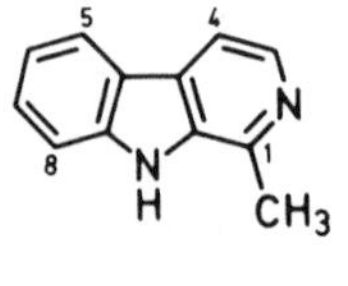

5/47, *Harman* 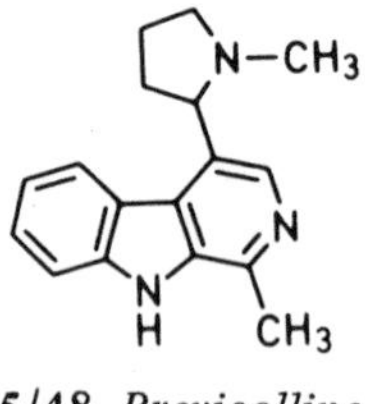   5/48, *Brevicolline*

A few bisalkaloids have β-carboline derivatives as one component (see Section 10).

5.1.2.3.2 Alkaloids of the Canthine Type Canthinone [5/49, from *Pentaceras* and *Xanthoxylum* spp. (RUT) and *Picrasma* sp. (SIM)] is a typical representative of this small group.

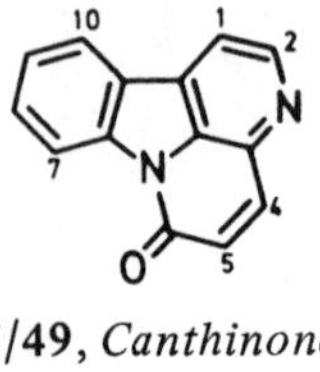

5/49, *Canthinone*

5.1.2.3.3 Eburnamine Alkaloids [T 7] , [T 8]   The structural variations typical of this subgroup are shown by eburnamine [5/50, from *Amsonia, Gonioma, Haplophyton, Hunteria, Pleiocarpa, Rhazya,* and *Vinca* spp. (APO)] , vincamine (5/51, from *Vinca* sp.), and schizozygine [5/52, from *Schizozygia coffeoides* (APO)] . A number of other naturally occurring bases are derived from these three alkaloids.

5/50, (−)-*Eburnamine*   5/51, (+)-*Vincamine*   5/52, (+)-*Schizozygine*

Representatives of this group also occur as one indole unit of certain bisindole alkaloids. (+)-Vincamine (5/51) is of pharmacological importance—it promotes the circulation of blood in the brain.

5.1.2.3.4 Heteroyohimbane and Biologically Related Bases [C 9] , [C 10]   This group of alkaloids can be further subdivided according to their carbon-nitrogen skeletons whose biogenetic relationships are well known:

Corynantheidine [5/53, from *Mitragyna* sp. (RUB) and *Pseudocinchona* sp. (RUB)]

Ajmalicine (5/54), an example of a heteroyohimbane alkaloid, from various plants of the genera *Catharanthus, Rauvolfia, Stemmadenia, Vinca,* and *Tonduzia* (APO), also *Corynanthe, Pausinystalia,* and *Mitragyna* (RUB)

Pleiocarpamine [5/55, from *Gonioma, Hunteria, Pleiocarpa,* and *Vallesia* spp. (APO); see Section 8.1]

Sarpagine [5/56, from *Rauvolfia* and *Vinca* spp. (APO)] [T 2] , [T 3]

Vobasine [5/57, from *Gabunia, Hazunta, Peschiera,* and *Voacanga* spp. (APO)]

Ajmaline[2] [5/58, from various plants of the genera *Aspidosperma, Rauvolfia,* and *Tonduzia* (APO)] [T 2] , [T 3]

Alstophylline [5/59, from *Alstonia* sp. (APO)]

Akuammiline [5/60, from *Conopharyngia* and *Picralima* spp. (APO)]

A few subtypes of these basic structures occur in which C-22 may or may not be present; rearranged products also occur naturally. Mitraphylline (5/61, from

---

[2] Ajmaline is employed in medicine to reduce cardiac irritability.

Mitragyna and *Vinca* spp.) is an oxindole derivative of ajmalicine (5/54), and oxindole derivatives are also known of alkaloids which belong to the coryanthei-dine (5/53) and vobasine (5/57) types. The pleiocarpamine (5/55) type is known to produce an indoxyl alkaloid: fluorocurine [5/62, from calabash curare and *Strychnos* sp. (STR); see Section 9.2].

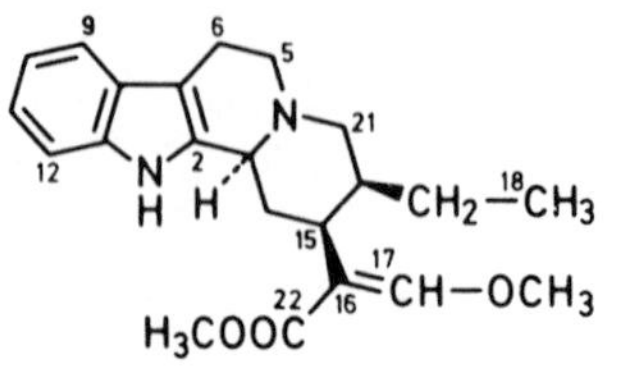

5/53, (−)-*Corynantheidine*

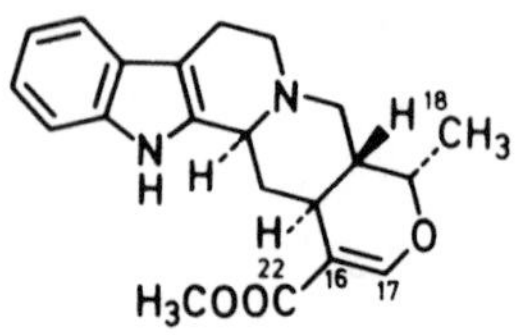

5/54, (−)-*Ajmalicine*

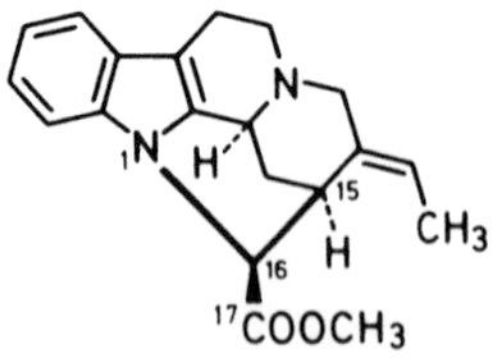

5/55, *Pleiocarpamine*

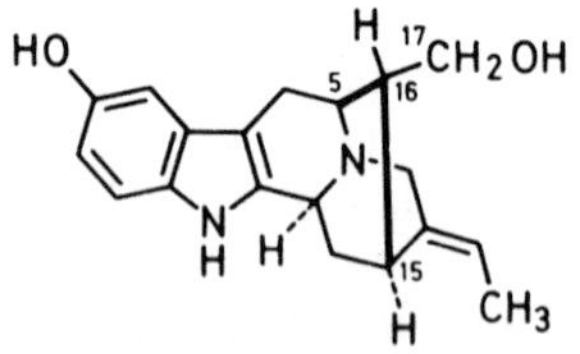

5/56, (+)-*Sarpagine*

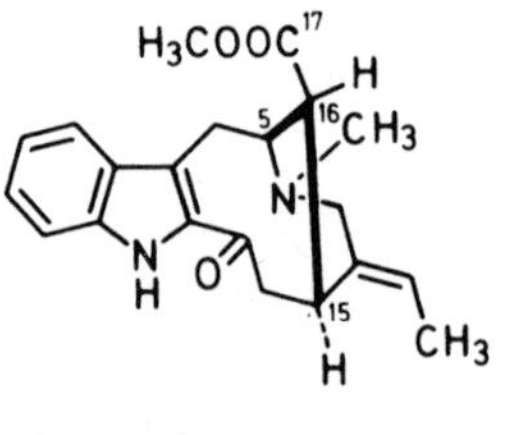

5/57, (−)-*Vobasine*

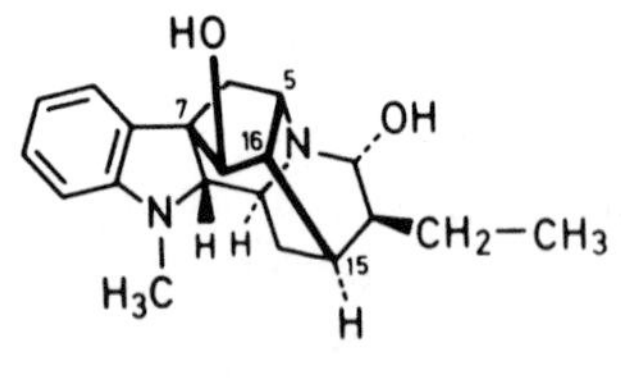

5/58, (+)-*Ajmaline*

The following alkaloids show further structural variations on the basic themes: alstonidine [5/53, from *Alstonia* sp. (APO)], indolopyridicoline [5/64, from *Gonioma* sp. (APO)], dehydrovoachalotine [5/65, from *Voacanga* sp. (APO)], picraphylline [5/66, from *Picralima* sp. (APO)], and aspidodasycarpine [5/67, from *Aspidosperma* sp. (APO)].

There are further alkaloid types that are closely related to the heteroyo-himbane alkaloids, of which vallesiachotamine [5/68, from *Vallesia* sp. (APO)], talbotine [5/69, from *Pleiocarpa* sp. (APO)], adifoline [5/70, from *Adina* sp. (RUB)], and cinchonamine [5/71, from *Remijia* sp. (RUB)] will serve as examples [T 4], [T 5], [T 6], [U 1].

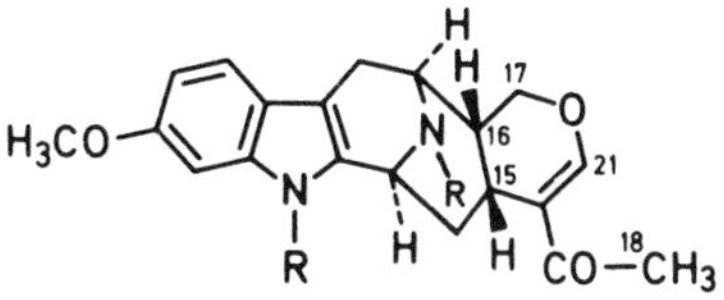

5/59, *Alstophylline*

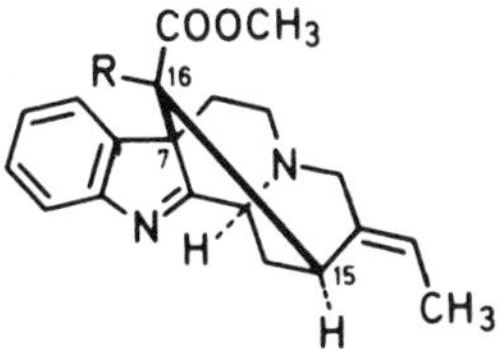

R=CH$_3$

5/60, (+)-*Akuammiline*

R = CH$_2$-O-CO-CH$_3$

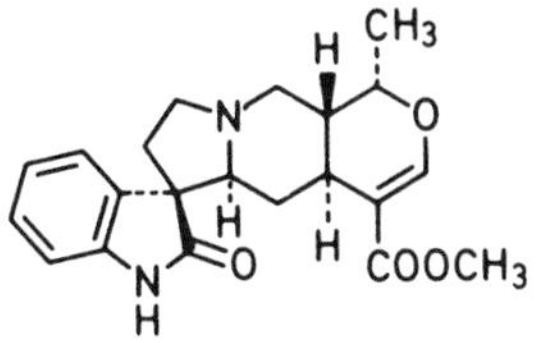

5/61, (−)-*Mitraphylline*

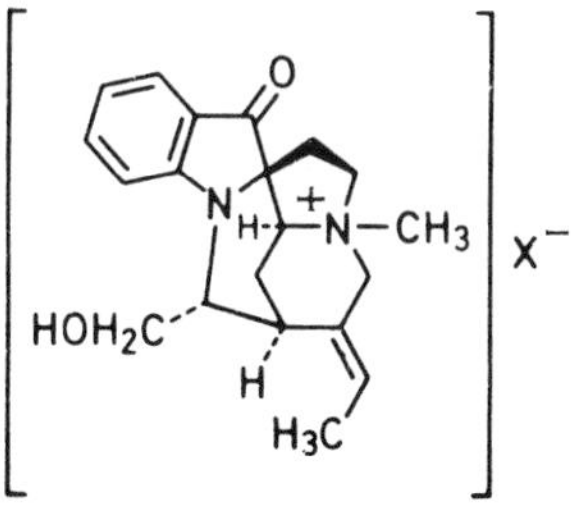

5/62, *Fluorocurine*

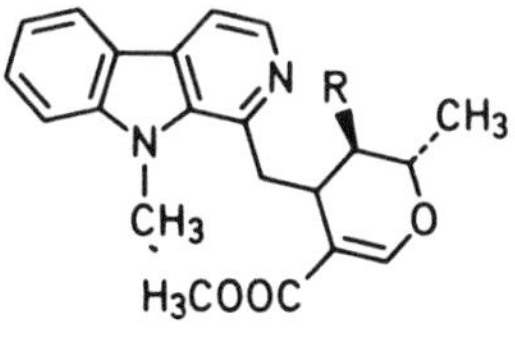

X$^-$

R = CH$_2$OH

5/63, (−)-*Alstonidine* 5/64, *Indolopyridicoline* 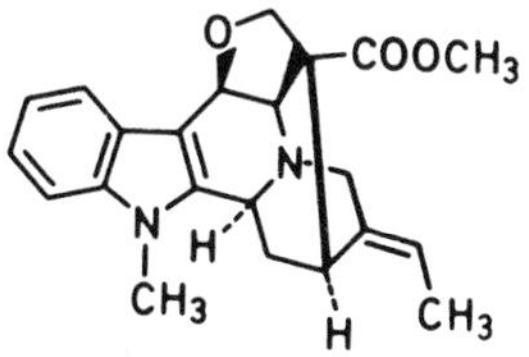 5/65, (+)-*Dehydrovoachalotine*

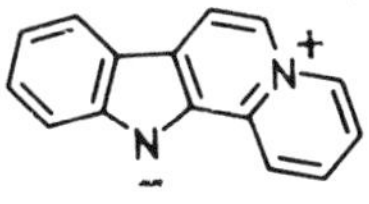

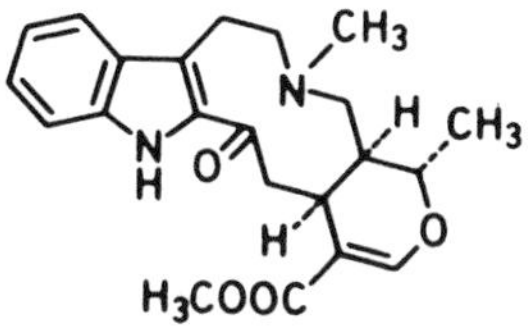

5/66, (−)-*Picraphylline*

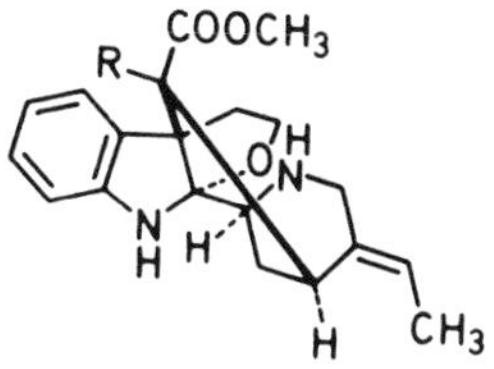

R = CH$_2$OH

5/67, (−)-*Aspidodasycarpine*

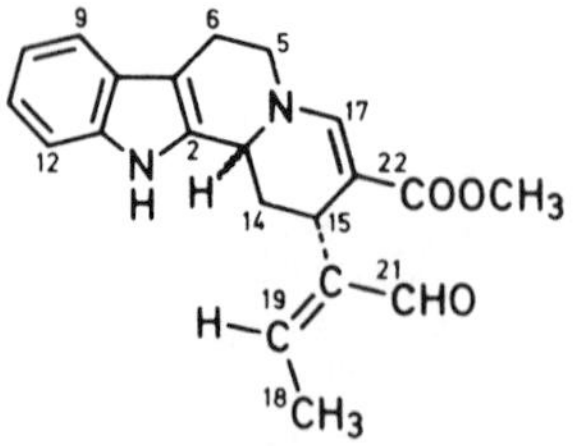

5/68, *Vallesiachotamine*

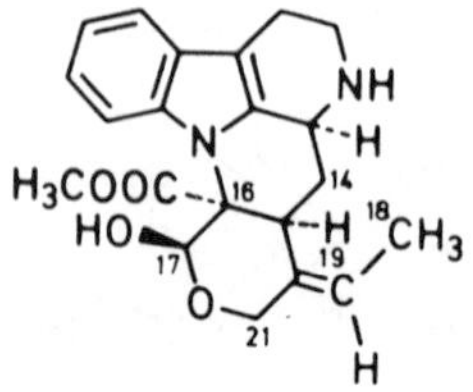

5/69, *Talbotine*

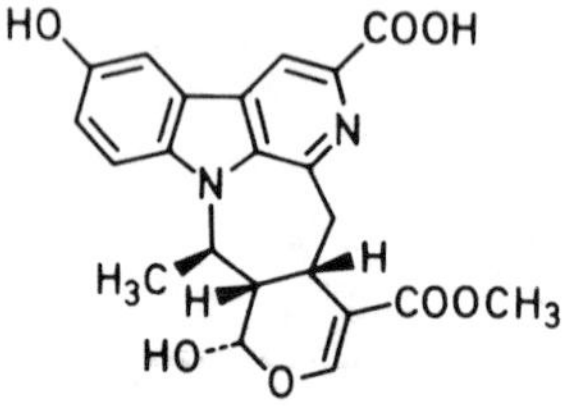

5/70, *Adifoline*

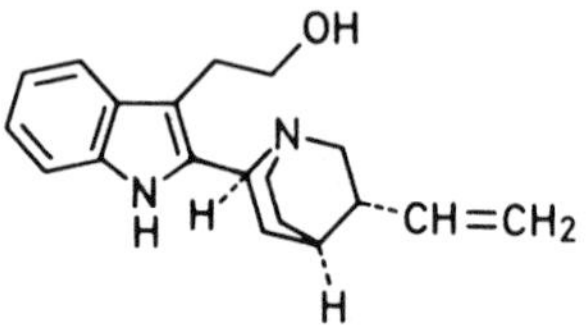

5/71, (+)-*Cinchonamine*

A few bases of the heteroyohimbane and related types form part of certain bisindole alkaloids (see Section 8).

5.1.2.3.5 Yohimbane Alkaloids [C 9], [C 10], [S 30] Two alkaloids are characteristic of this group: yohimbine [5/72, from plants of the genera *Alchornea* (EUP); *Aspidosperma*, *Catharanthus*, *Diplorrhynchus*, *Rauvolfia*, *Vinca* (APO); *Corynanthe* and *Pausinystalia* (RUB)] and reserpine [5/73, from various species of *Alstonia*, *Excavatia*, *Ochrosia*, *Rauvolfia*, *Tonduzia*, *Vallesia*, and *Vinca* (APO)]. Various other bases that are stereoisomers, structural

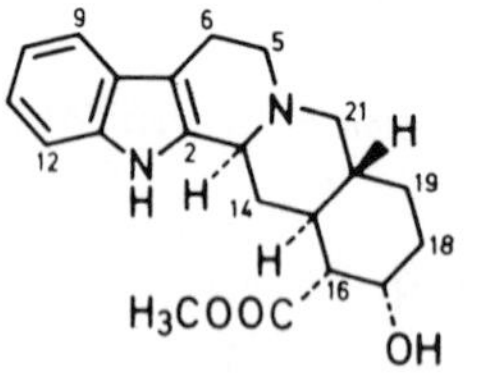

5/72, (+)-*Yohimbine*

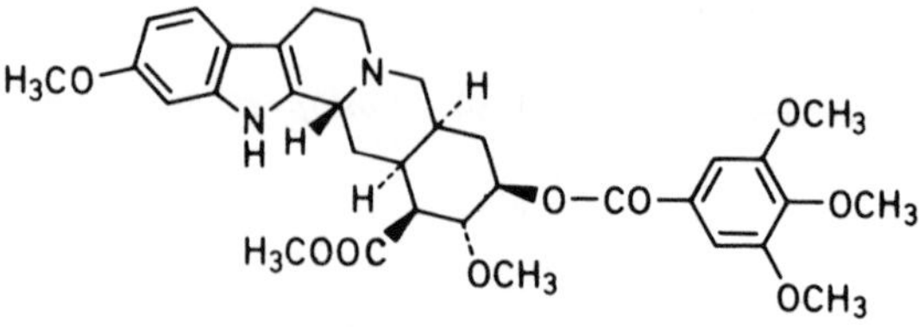

5/73, (−)-*Reserpine*

isomers, or derivatives of these occur naturally. Alstonilidine [5/74, from *Alstonia* sp. (APO)] may be regarded as a derivative of this type of alkaloid. Yohimbine is an aphrodisiac (in veterinary medicine), while reserpine and certain of its derivatives are used in treating hypotension.

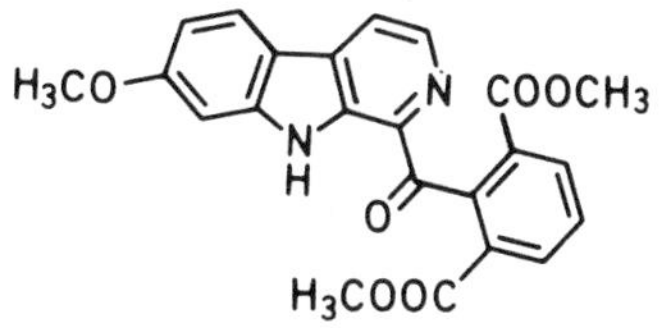

5/74, *Alstonilidine*

5.1.2.4 Alkaloids with a γ-Carboline Skeleton [S 5]

The only alkaloid so far known to belong to this group is cryptolepine [5/75, from *Cryptolepis* sp. (PER)].

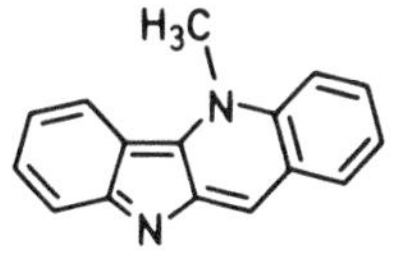

5/75, *Cryptolepine*

5.1.2.5 Iboga Alkaloids [T 13], [T 14]

In this group of alkaloids the indole nucleus is fused onto a seven-membered ring containing another nitrogen. The alkaloids ibogamine [5/76, from *Gabunia*, *Hazunta*, *Stemmadenia*, *Tabernaemontana*, and *Tabernanthe* spp. (APO)] and voacangine [5/77, from *Callichilia*, *Conopharyngia*, *Gabunia*, *Hedranthera*, *Peschiera*, *Rejoua*, *Stemmadenia*, *Tabernaemontana*, and *Voacanga* spp. (APO)] are typical of this class. Various oxidation products occur also as in the case of the heteroyohimbane bases. Other representatives include hydroxyindolenine-

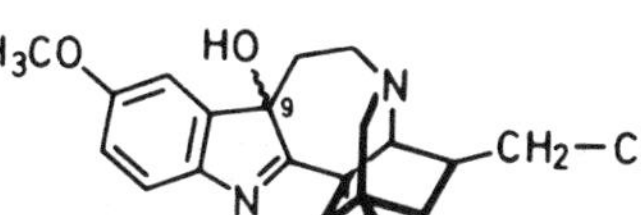

5/76, (−)-*Ibogamine*

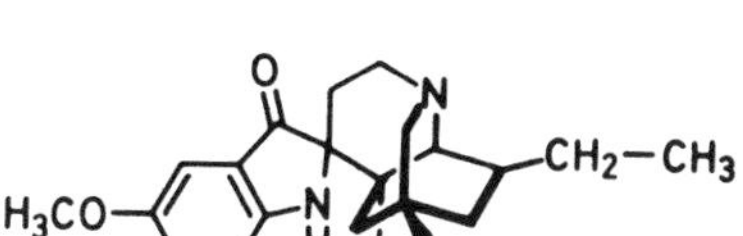

5/77, (−)-*Voacangine*

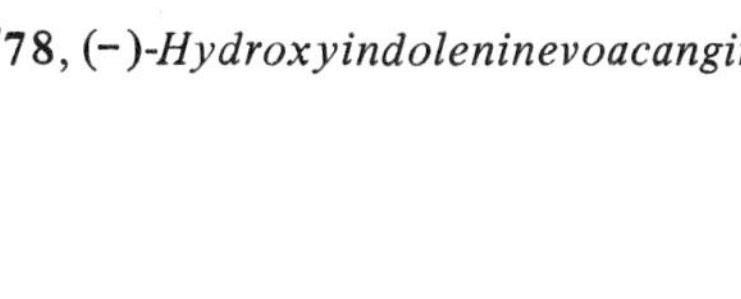

5/78, (−)-*Hydroxyindoleninevoacangine*

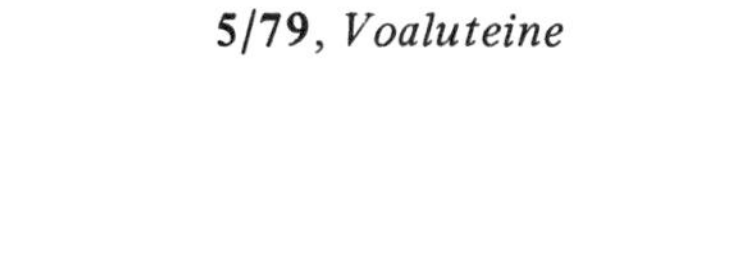

5/79, *Voaluteine*

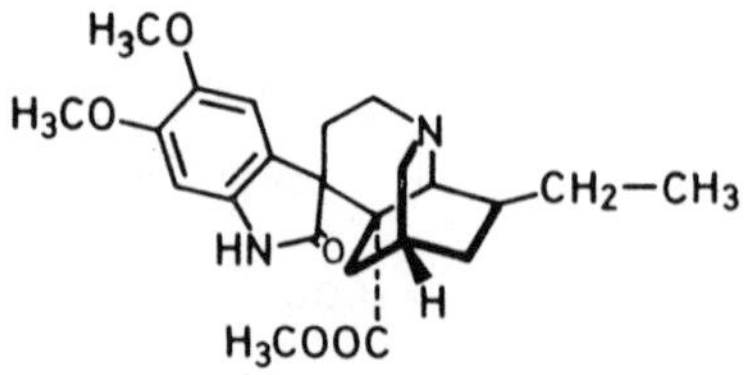

5/80, *Crassanine*

voacangine [5/78, from *Voacanga* sp. (APO)], voaluteine [5/79, from *Rejoua* sp. (APO)], and crassanine [5/80, from *Tabernaemontana* sp. (APO)]. In addition to these there is also a range of alkaloids with a different substitution pattern. Bisindole alkaloids in which one portion is an *Iboga* alkaloid have also been isolated from the Apocynaceae (see Section 10.6).

5.1.2.6 Indole Alkaloids with a Pyrrolidinoindole Skeleton

5.1.2.6.1 Eserine-Type Alkaloids [C 4], [R 4], [R 11] So far two structural variants of the eserine alkaloids are known: physostigmine (eserine, 5/81) from *Dioclea* and *Physostigma* spp. (LEG) and *Hippomane* sp. (EUP); and geneserine (5/82, from *Physostigma* sp.).

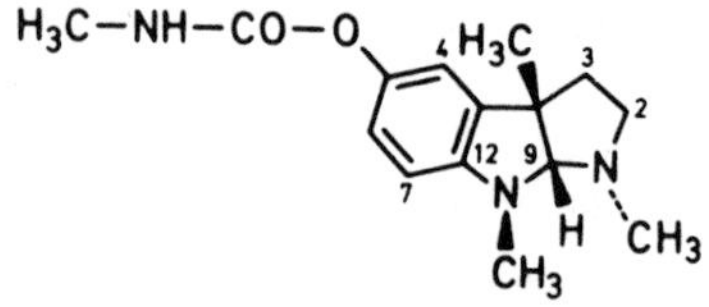

5/81, (−)-*Physostigmine (Eserine)* 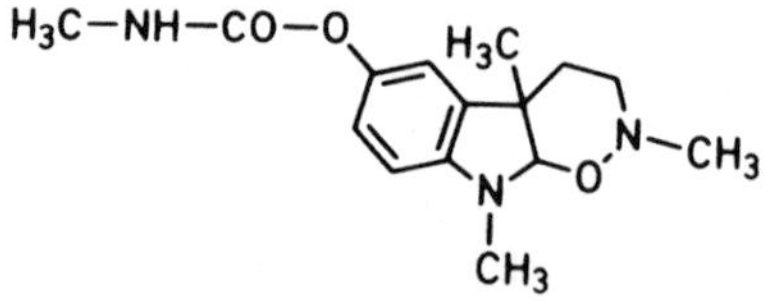   **5/82**, *Geneserine*

Various bases from *Calycanthus* spp. (CAL) [M 5] with an indole nucleus also belong to this group, but since they occur as di-, tri-, or tetrameric compounds and no monomeric representative is yet known, they have been classified with the bisalkaloids (see Section 10.4).

5.1.2.6.2 Echitamine-Erinine-Type Alkaloids There is a clear biogenetic relationship between the alkaloids echitamine [5/83, from *Alstonia* sp. (APO)], corymine [5/84, from *Hunteria* sp. (APO)], and erinine [5/85, from *Hunteria* sp. (APO)] on the one hand, and the heteroyohimbane alkaloids (see Section 5.1.2.3.4) on the other, but in accord with the principle of classification that we have adopted, they are grouped under this heading.

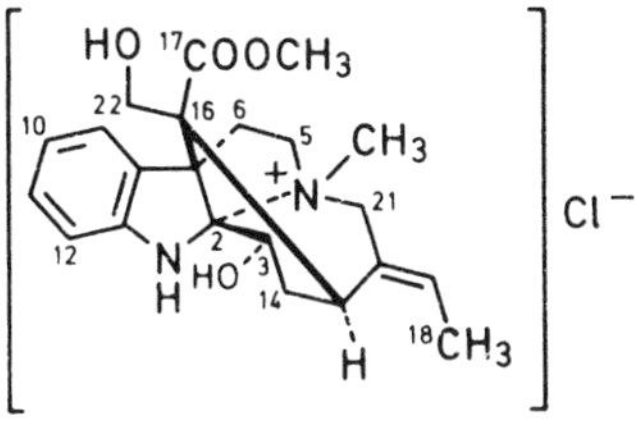

5/83, (−)-Echitamine chloride

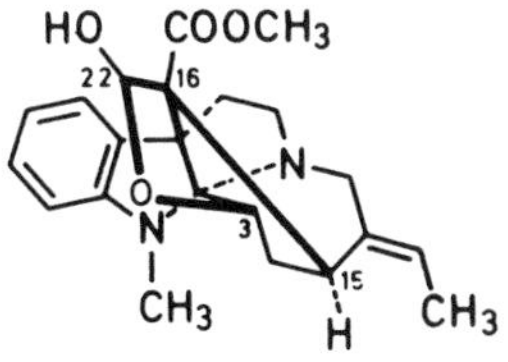

5/84, (+)-Corymine

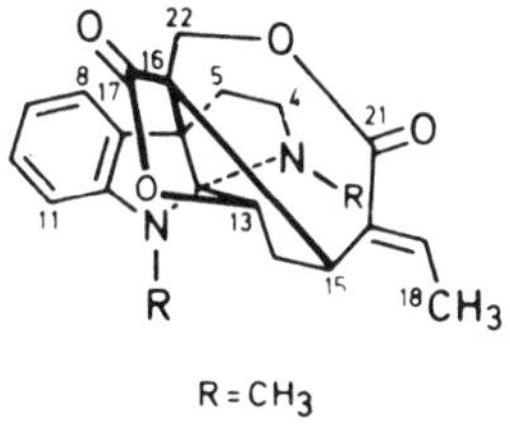

R = CH₃

5/85, (−)-Erinine

5.1.2.7 Ergot Alkaloids [G 5], [S 19], [S 20], [S 32], [S 50]

These alkaloids from *Claviceps purpurea* have lysergic acid (5/86) or isolysergic acid (8-*epi*-5/86) as the fundamental building block. Two further basic types, agroclavine (5/87) and chanoclavine (5/88), are derived from 5/86. Apart from simple amides like ergobasine (5/89), peptide alkaloids such as ergotamine

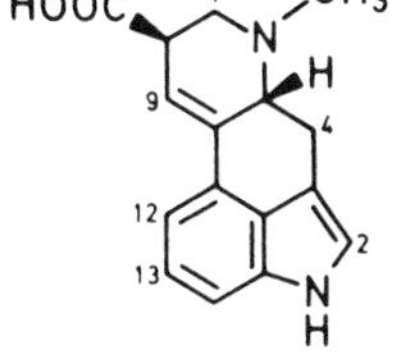

5/86, *Lysergic acid*

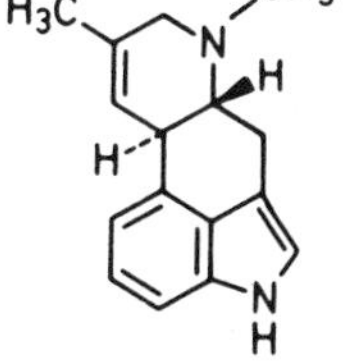

5/87, *Agroclavine*

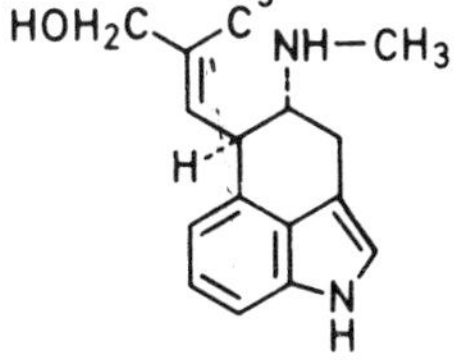

5/88, *Chanoclavine*

(5/90) and ergocornine (5/91) also occur. Ergot alkaloids have been successfully used in various branches of medicine (e.g., obstetrics, neurology). Lysergic acid diethylamide (LSD) has structure 5/92.

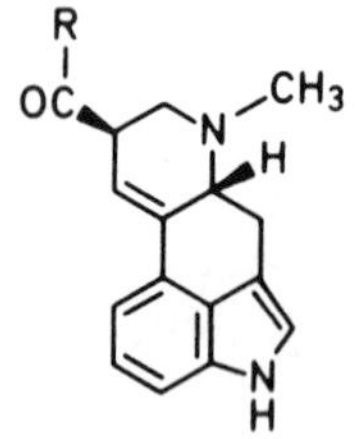

$R = NH-\overset{\underset{\textstyle |}{CH_3}}{CH}-CH_2OH$

5/89, *Ergobasine*

$R = N(C_2H_5)_2$

5/92, LSD

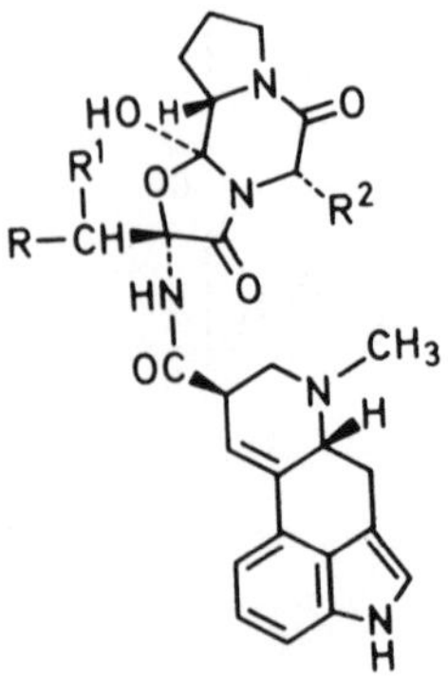

$R = R^1 = H, R^2 = CH_2\text{-}C_6H_5$

5/90, *Ergotamine*

$R = R^1 = CH_3, R^2 = CH(CH_3)_2$

5/91, *Ergocornine*

5.1.2.8 Evodiamine-Type Alkaloids [M 7]

Evodiamine [**5/93**, from *Evodia* sp. (RUT)] is the prototype of this relatively small group of alkaloids, which may also be considered to be quinazoline alkaloids.

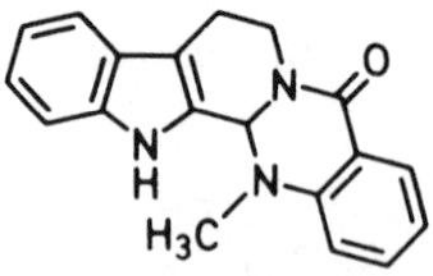

5/93, *Evodiamine*

5.1.3 Piperidine Alkaloids [G 19], [H 5]

This group has the nitrogen in a six-membered ring. The simplest piperidine bases resemble the corresponding pyrrolidine derivatives. Thus piperine (**5/94**) from *Piper* sp. (PIP) may be compared with **5/1**. An example of an α-alkyl-piperidine is provided by coniine [**5/95**, from *Aethusa* and *Conium* spp. (UMB)]. Anaferine [**5/96**, from *Withania* sp. (SOL)] is an example of a compound with two piperidine nuclei. Other examples include lobeline [**5/97**, from *Lobelia*

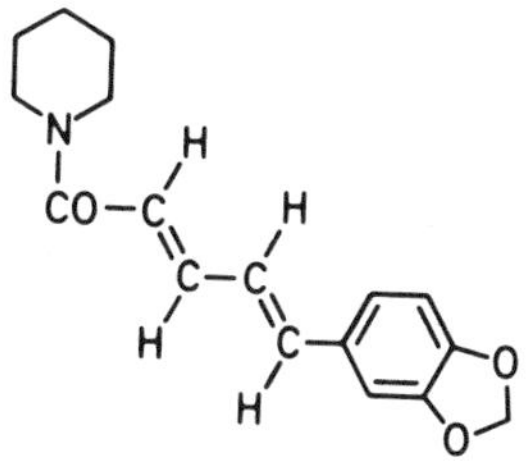

5/94, *Piperine*

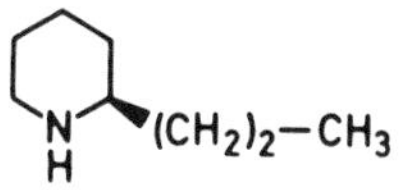

5/95, (−)-*Coniine*

sp. (CAM)] and the skytanthines, such as α-skytanthine [5/98, from *Skytanthus* sp. (APO)], which consist of a monoterpenoid unit and methylamine.

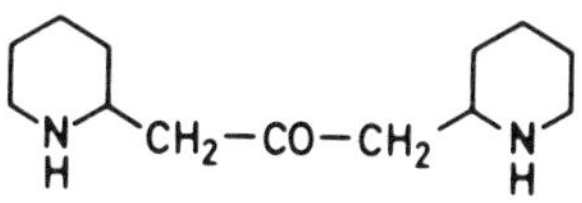

5/96, *Anaferine*

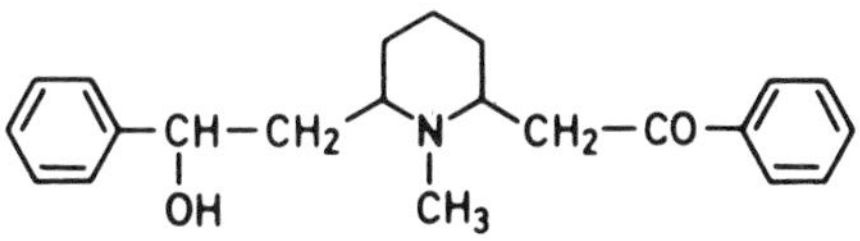

5/97, *Lobeline*

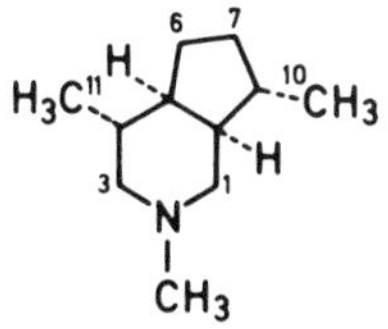

5/98, α-*Skytanthine*

5.1.4 Pyridine Alkaloids [A 2], [G 9], [H 5], [M 25], [M 26]

The pyridine nucleus occurs in cantleyine [5/99, from *Cantleya* sp. (ICA) and *Jasminum* sp. (OLE)]. The skeleton is the same as in skytanthine (5/98), but in a dehydrogenated form. Both types may be classified as terpene alkaloids since their carbon skeleton arises from a monoterpene. Ricinine (2/3) and trigonelline [5/100, widely distributed, e.g., from *Trigonella* sp. (LEG)] may be cited as examples of simple pyridine alkaloids. The latter is a nicotinic acid derivative and clearly forms a borderline case as far as alkaloids are concerned. Evonine [5/101, from *Evonymus* sp. (CEL)], which contains two ester groups, belongs to this structural type.

A group of bases referred to as tobacco alkaloids occupy an intermediate position between the pyrrolidine, piperidine, and pyridine bases. They may be represented here by nicotine [5/102, from several families, e.g., from *Nicotiana tabacum* (SOL)] and anabasine (5/103, especially common in *Nicotiana* spp.). (−)-Nicotine stimulates the central nervous system in small doses; larger doses lead to central paralysis. It is obtained commercially from tobacco wastes. Bases of the halfordine type [5/104, from *Halfordia* sp. (RUT)] are related to the tobacco alkaloids.

5/99, *Cantleyine*   5/100, *Trigonelline*   5/101, *Evonine*

$$R^1 = CH_3$$
$$R^2 = O\text{-}\text{-}CO\text{-}CH_3$$

5/102, (−)-*Nicotine*   5/103, *Anabasine*   5/104, *Halfordine*

5.1.5 Tropane and Related Bases [F 1], [F 2], [F 3], [F 6], [F 8], [H 6]

A piperidine ring is also present in the bicyclic alkaloid systems of tropinone [5/105, from *Nicandra* sp. (SOL)] and pseudopelletierine [5/106, from *Punica*

5/105, *Tropinone*   5/106, ψ-*Pelletierine*

sp. (PUN)]. A large number of alkaloids, many of pharmacological importance, have the tropinone nucleus: for example, atropine [from *Atropa belladona* (SOL)] dilates the pupil of the eye; cocaine [from the leaves of the South American bush *Erythroxylum coca* (ERY)] has a local anaesthetic effect and is used to produce euphoria and excitement; hyoscyamine [widely distributed, e.g., in *Hyoscyamus* sp. (SOL)]; scopolamine [widely distributed, e.g., in *Scopolia* sp. (SOL)] has an action similar to that of atropine and depresses the central nervous system. In large doses these alkaloids are highly toxic.

5.1.6   Histamine, Imidazole, and Guanidine Alkaloids [B 19]

This relatively small group of bases generally has a five- or six-membered ring with two nitrogen atoms. The alkaloid glochidine [5/107, from *Glochidion* sp. (EUP)] is an example of a histamine derivative, while alchornine [5/108, from *Alchornea* sp. (EUP)] may be regarded as an imidazole-pyrimidine derivative.

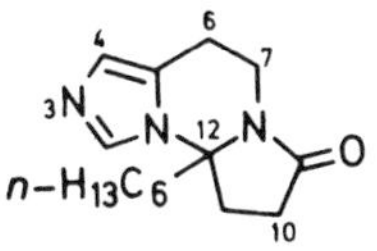

5/107, *Glochidine* 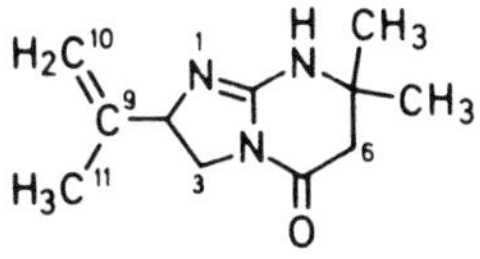   5/108, *Alchornine*

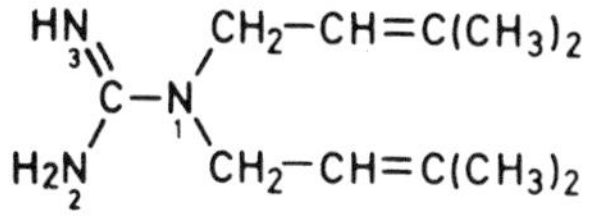

5/109, N^1,N^1-*Diisopentenylguanidine*

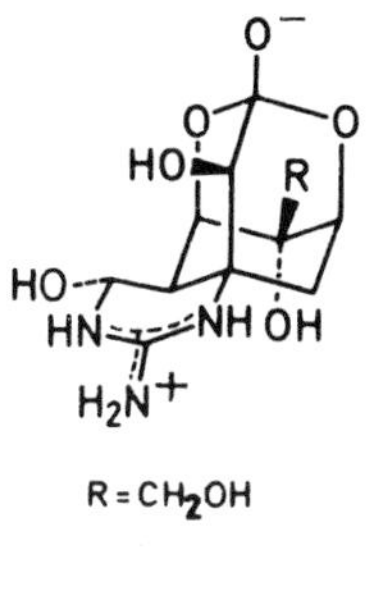

R = CH_2OH

5/110, *Tetrodotoxin*

There is a very small number of naturally occurring bases belonging to this alkaloid type with a guanidine residue incorporated into the nucleus. Examples are provided by N',N'-diisopentenylguanidine [5/109, from *Alchornea* sp. (EUP)] and the highly toxic tetrodotoxin (5/110, e.g., from the puffer fish, *Fugu rubripes rubripes*).

5.1.7 Isoquinoline Alkaloids

Bibliography

General reviews: [K 6] , [M 12] , [S 15] , [S 38]
Special reviews: [G 3] , [T 12]
Structure: [K 6]
Biogenesis: [M 13]
Papaveraceous alkaloids: [M 27] , [P 18] , [S 26]
Thalictrum alkaloids: [M 42]

The total number of isoquinoline alkaloids is at present about 700 and is thus comparable with the number of indole alkaloids. On biogenetic grounds various plant bases are included in this group which have no isoquinoline residue, such as rhoeadine, papaverrubine, and protopine alkaloids.

5.1.7.1 Simple Isoquinoline Alkaloids [B 2], [R 6], [R 7]

These alkaloids are derived from tetrahydroisoquinoline and for the most part have a carbon chain attached to C-1, often a one-carbon substituent. Typical representatives are hydrohydrastinine [5/111, from *Corydalis* sp. (FUM)], salsoline [5/112, from *Salsola* sp. (CHE)], and lophocerine [5/113, from *Lophocereus* sp. (CAC)]. The group also includes lactam alkaloids (C=O at C-1), and another member is the tricyclic alkaloid peyoglutam (5/114). The interesting alkaloid ancistrocladine [5/115, from *Ancistrocladus* sp. (ANC)] provides a novel type of skeleton.

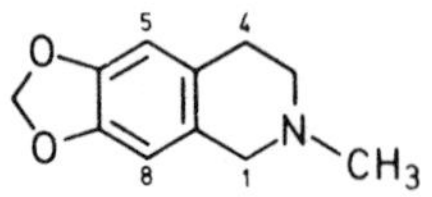

5/111, *Hydrohydrastinine* 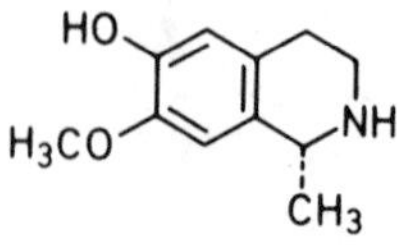   5/112, (+)-*Salsoline* 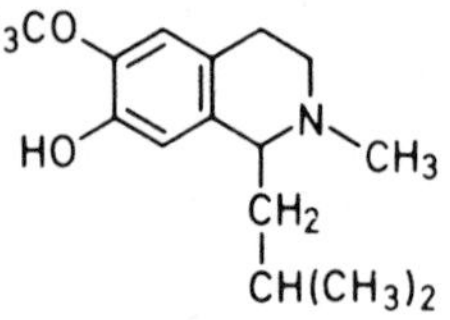   5/113, *Lophocerine*

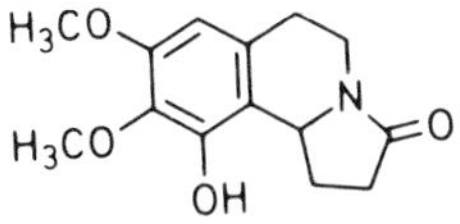

5/114, *Peyoglutam*

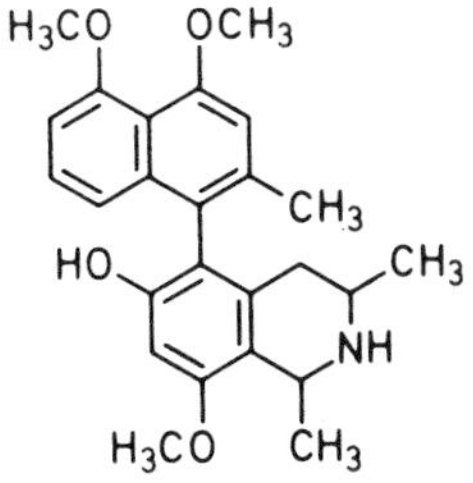

5/115, *Ancistrocladine*

5.1.7.2 Benzylisoquinoline Alkaloids [B 3], [D 1], [K 8] (see Section 6)

In a few alkaloids of this class, ring B is aromatic [e.g., papaverine (5/116) from *Papaver* sp. (PAP)], but they are mostly derivatives of tetrahydroisoquinoline, such as laudanosine [5/117, from *Papaver* sp. (PAP)]. Papaverine has a general relaxing effect on the smooth musculature, and is commonly used in medicine for treating conditions of cramp.

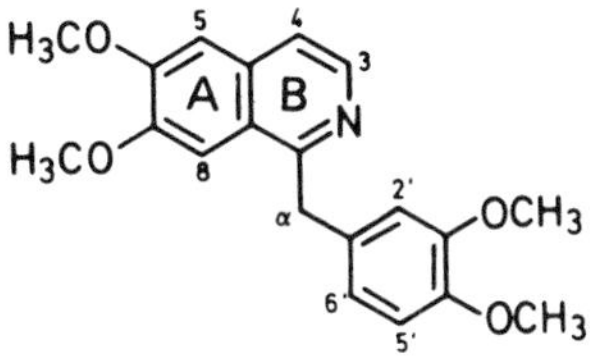

5/116, *Papaverine*

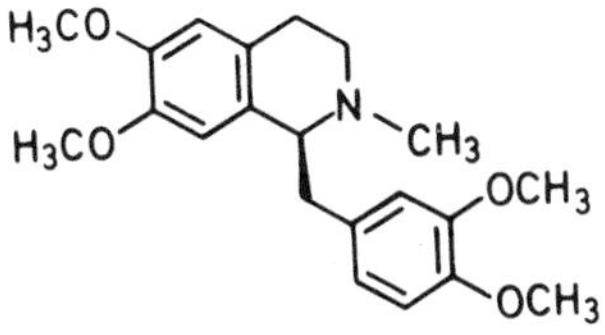

5/117, (+)-*Laudanosine*

A number of quaternary ammonium derivatives occur naturally in this as well as in other alkaloid series.

An example of an *N*-benzylisoquinoline is sendaverine [5/118, from *Corydalis* sp (FUM)]. The so-called bisbenzylisoquinoline alkaloids occur in nature more frequently than the monomeric bases (see Section 10.4)

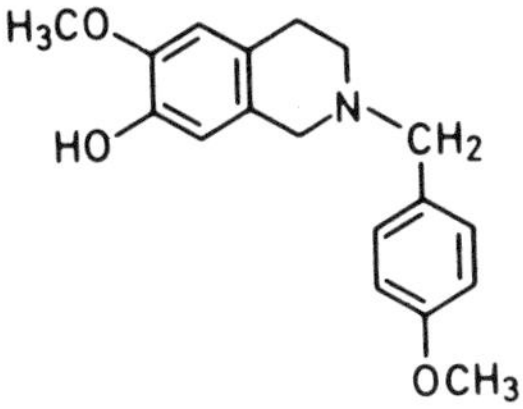

5/118, *Sendaverine*

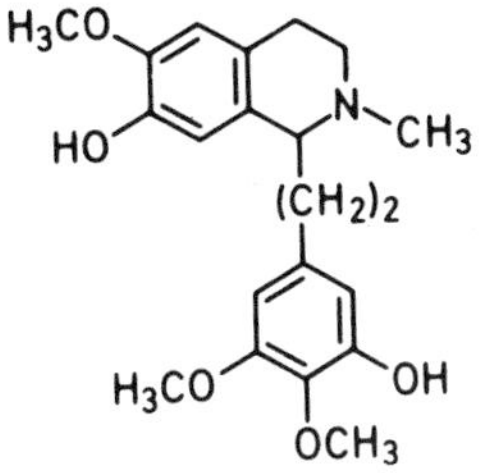

5/119, *Autumnaline*

Homologues of the benzylisoquinoline alkaloids are also known, such as autumnaline [5/119, from *Colchicum* sp. (LIL)] .

5.1.7.3 Phenyltetrahydroisoquinoline Alkaloids

This group of alkaloids, which so far has only a few members, differs from the benzylisoquinoline alkaloids in having no methylene group between the isoquinoline residue and the benzene ring. Cryptostyline I [5/120, from *Cryptostylis* sp. (ORC)] provides an example.

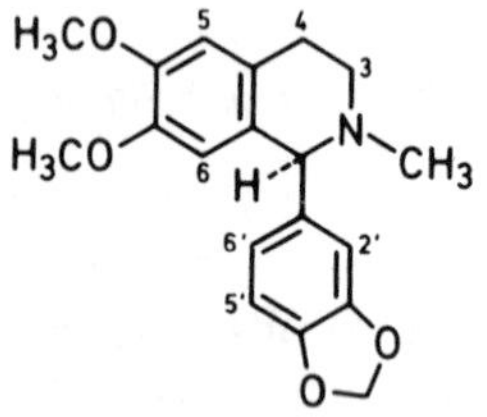

5/120, (+)-*Cryptostyline I* 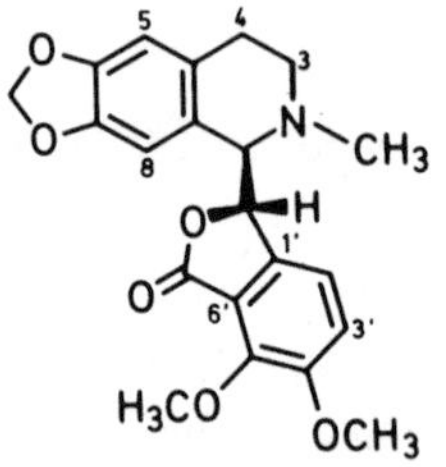   5/121, (–)-*Hydrastine*

5.1.7.4 Phthalideisoquinoline Alkaloids [S 23], [S 24], [S 25] (See Section 6)

A typical example of this group is hydrastine [5/121, from *Berberis* sp. (BER) and *Hydrastis* sp. (HYD)] . (–)-α-Narcotine, used as a pectoral, also belongs to this small group of alkaloids.

5.1.7.5 Rhoeadine and Papaverrubine Alkaloids (See Sections 6 and 9.1.2)

Rhoeadine [5/122, from *Papaver* sp. (PAP)] and papaverrubine A (5/123, from *Papaver* sp.) are characteristic of this subgroup.

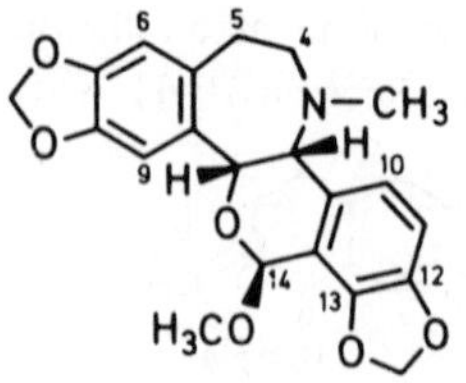

5/122, (+)-*Rhoeadine* 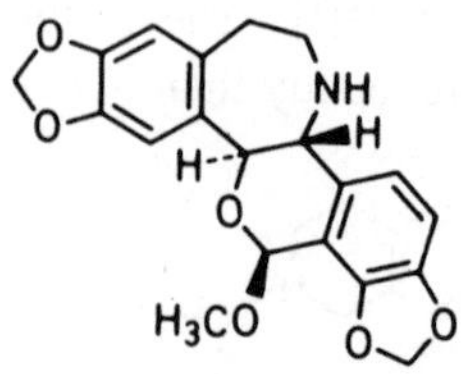   5/123, (+)-*Papaverrubine A*

5.1.7.6 Ipecacuanha Alkaloids [B 34], [J 1], [M 15], [O 1], [S 56]

This group comprises compounds such as protoemetine [5/124, from *Psychotria* and *Uragoga* spp. (RUB)] and ipecoside (5/125, from *Psychotria* sp.) as well as alkaloids with two nitrogen atoms, such as emetine [5/126, from *Alangium* sp. (ALA) and *Borreria, Bothriospora, Capirona, Ferdinandusa, Hillia, Psychotria, Remijia*, and *Tocoyena* spp. (RUB)] . Emetine is active against amebic dysentery. A further group of bases that are closely related to the ipecacuanha alkaloids are dealt with under bisalkaloids (see Section 10.3). There is an obvious biogenetic relationship between ipecoside (5/125) and the heteroyohimbane alkaloids.

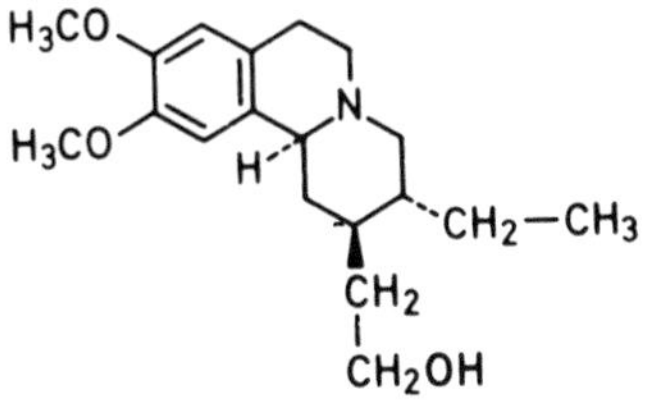

5/124, (−)-*Protoemetine*

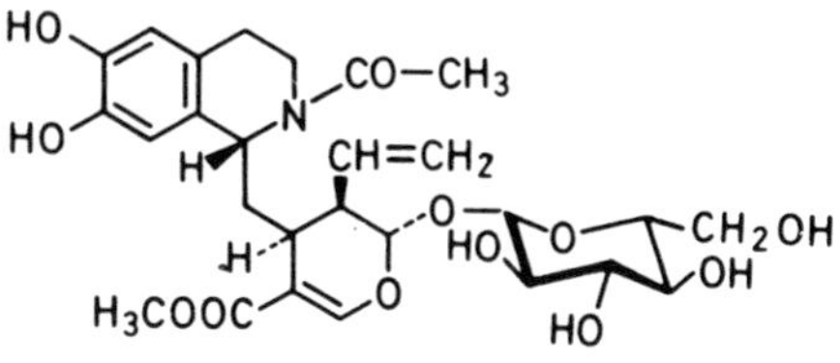

5/125, (−)-*Ipecoside*

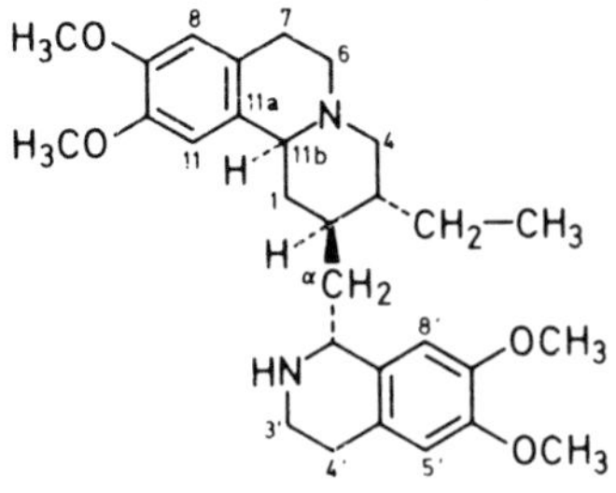

5/126, (−)-*Emetine*

5.1.7.7 Cryptaustoline-Type Alkaloids

An example of this very small group of alkaloids is provided by cryptaustoline [5/127, from *Cryptocarya* sp. (LAU)] (see Section 9.2).

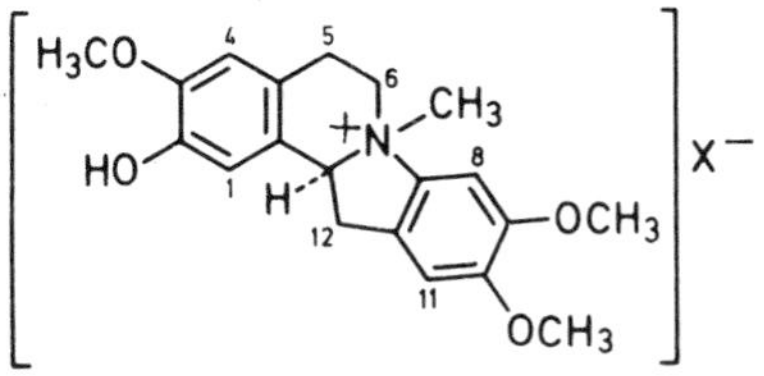

5/127, (−)-*Cryptaustoline*

5.1.7.8 Aporphine and Homoaporphine Alkaloids [M 2], [S 2], [S 36], [S 54] (See Section 6)

Glaucine [5/128, from *Corydalis* and *Dicentra* spp. (FUM) and *Glaucium* sp. (PAP)] and corydine (5/129, from *Corydalis*, *Dicentra*, and *Glaucium* spp.) are characteristic examples of the aporphine alkaloids. Examples of oxoaporphines are also known, such as liriodenine [5/130, from *Asimina* sp. (ANN), *Atherosperma* and *Doryphora* spp. (ATH), *Lysichitum* sp. (ARA), and from *Liriodendron*, *Magnolia*, and *Michelia* spp. (MAG)]. Thalphenine (5/131), isolated from *Thalictrum* sp. (RAN), also belongs to this group. In the homoaporphine alkaloids, ring C is seven-membered, such as multifloramine [5/132, from *Kreysigia* sp. (LIL)].

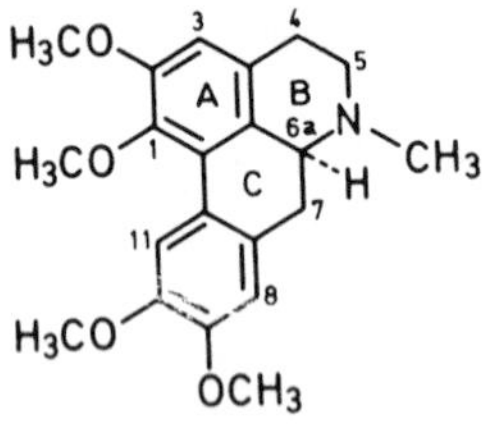

5/128, (+)-*Glaucine* 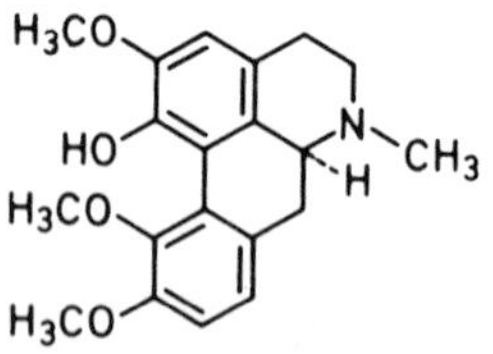   5/129, (+)-*Corydine* 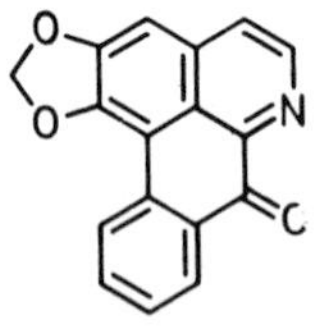   5/130, *Liriodenine*

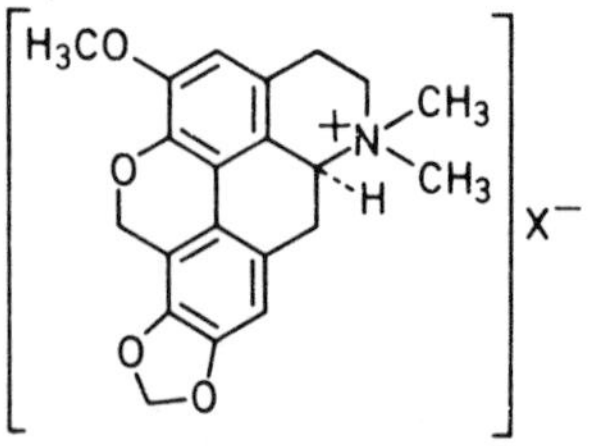

5/131, (+)-*Thalphenine* 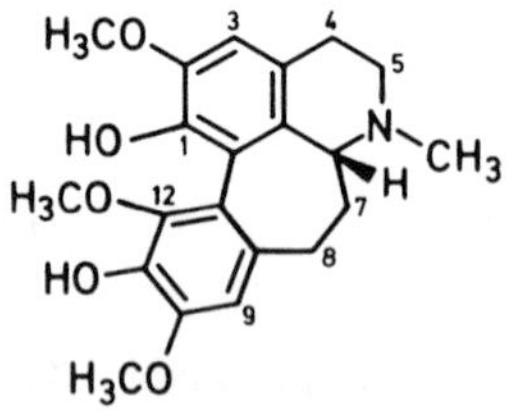   5/132, (−)-*Multifloramine*

5.1.7.9 Proaporphine and Homoproaporphine Alkaloids [B 13], [S 37] (See Section 6)

A typical example of this class is pronuciferine [5/133, from various species of *Croton* (EUP), *Nelumbo* (NEL), *Papaver* (PAP), and *Stephania* (MEN)]. Examples are also known of alkaloids with a different oxygen substitution pattern, and others with ring D reduced. In the homoproaporphine alkaloids, for example, bulbocodine [5/134, from *Bulbocodium* sp. (LIL)], ring C has been enlarged by one carbon atom.

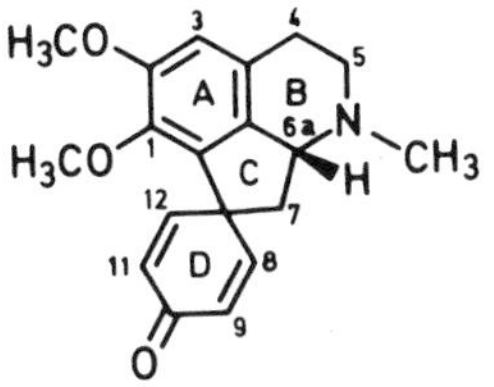

5/133, (+)-*Pronuciferine*

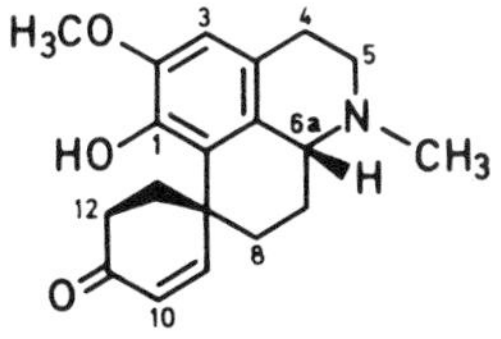

5/134, (+)-*Bulbocodine*

5.1.7.10 Cularine Alkaloids [M 31/32]

Cularine [5/135, from *Corydalis* sp. (FUM)] is an example of this very small group of isoquinoline alkaloids (see also bisalkaloids, Section 10.6).

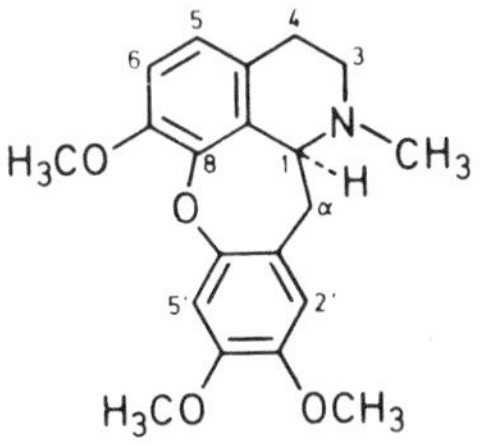

5/135, (+)-*Cularine*

5.1.7.11 Berberine Alkaloids [J 2], [M 22] (See Section 6)

In addition to alkaloids of the type of canadine (5/136) from *Corydalis* (FUM), *Hydrastis* (HYD), and *Xanthoxylum* (RUT) spp., and derivatives in which ring C has been dehydrogenated, such as berberine (5/137), examples also occur naturally that have undergone oxidation, for example, ophiocarpine [5/138, from *Corydalis* sp. (FUM)], and others with an extra carbon substituent, such as mecambridine [5/139, from *Papaver* sp. (PAP)] and corydaline (5/140, from *Corydalis* sp.).

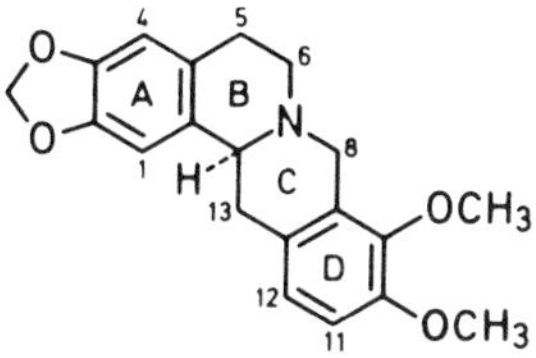

5/136, (−)-*Canadine*

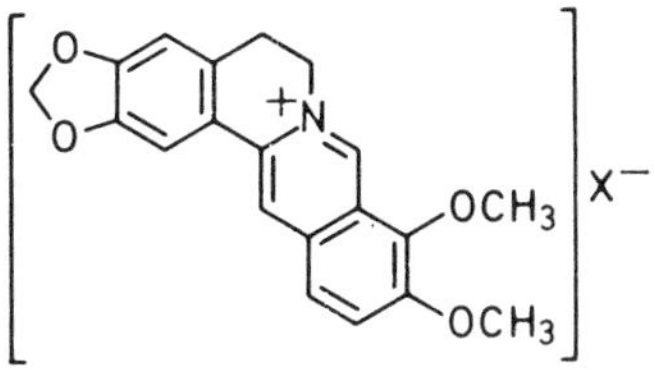

5/137, *Berberine*

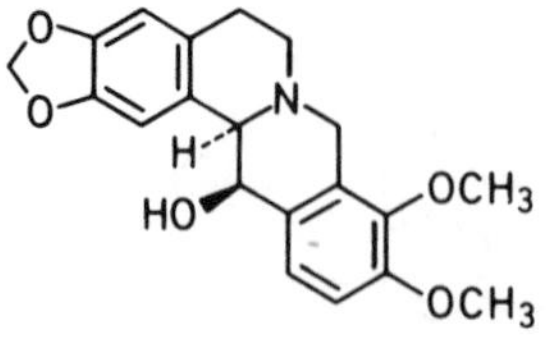

5/138, (−)-*Ophiocarpine*

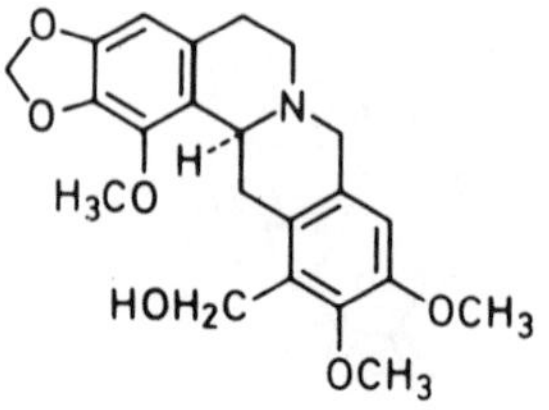

5/139, (−)-*Mecambridine*

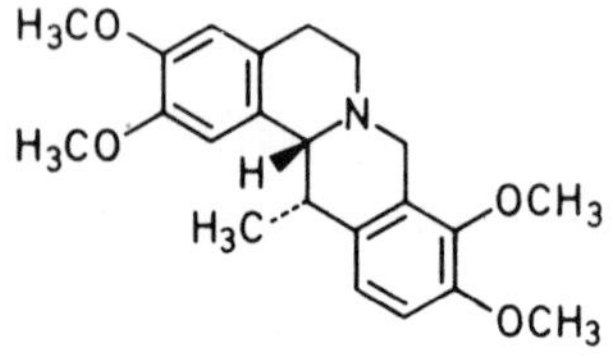

5/140, (+)-*Corydaline*

5.1.7.12   Protopine Alkaloids [M 21] (See Section 6)

The chief alkaloid of this group is protopine (5/141, widely distributed in the families FUM, HYP, NAN, PAP, and PTE), but alkaloids of the type of oxomuramine [5/142, from *Papaver* sp.(PAP)] and corycavidine [5/143, from *Corydalis* sp. (FUM)] also occur. All three alkaloids show structural similarities to certain bases of the berberine type.

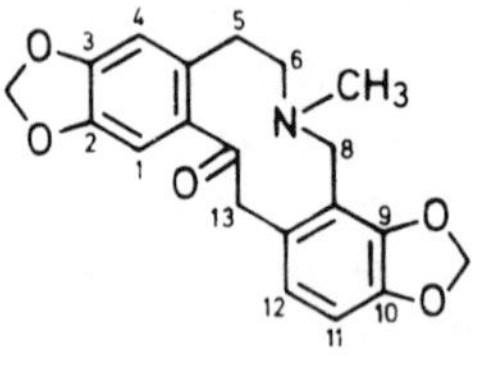

5/141, *Protopine*

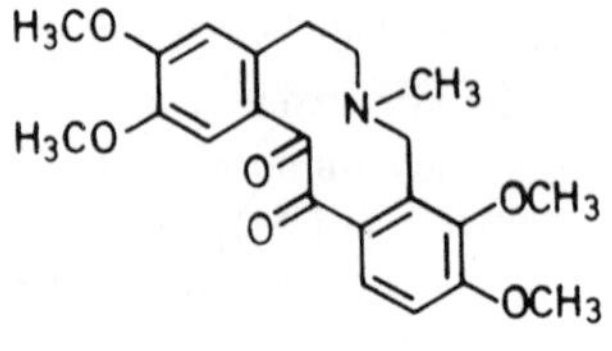

5/142, *Oxomuramine*

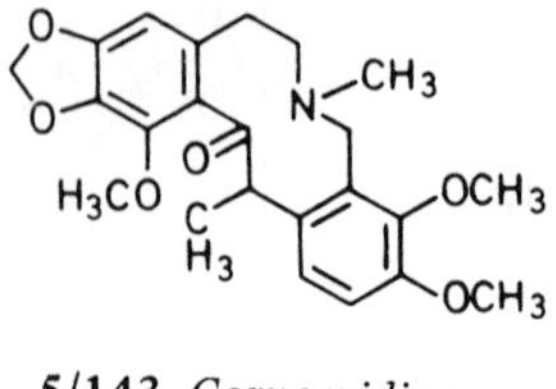

5/143, *Corycavidine*

5.1.7.13 Benzophenanthridine Alkaloids [M 19/20] (See Section 6)

Typical alkaloids of this group are sanguinarine (5/144, widely distributed in the families FUM, HYP, PAP, and RUT), oxynitidine [5/145, from *Xanthoxylum* sp. (RUT)], corynolamine [5/146, from *Corydalis* sp. (FUM)], and corynoloxine (5/147, from *Corydalis* sp.). Benzophenanthridine derivatives can dimerize to form bisalkaloids (see Section 10.1).

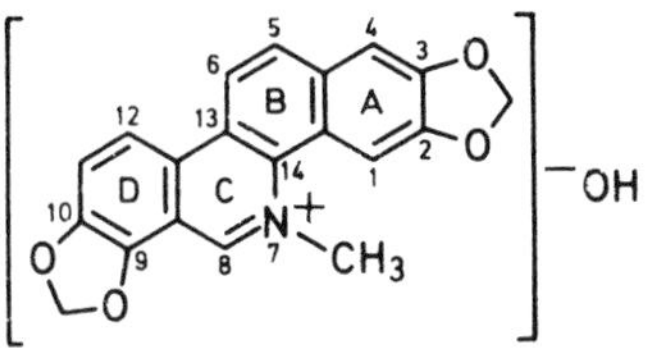

5/144, *Sanguinarine* 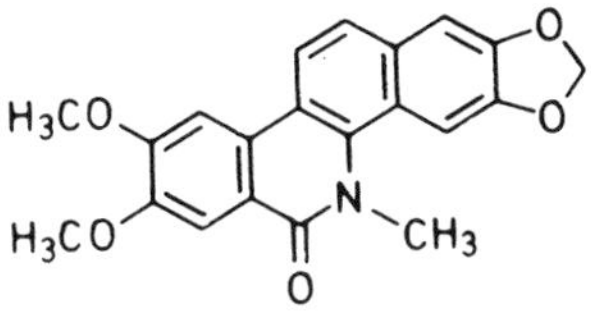   5/145, *Oxynitidine*

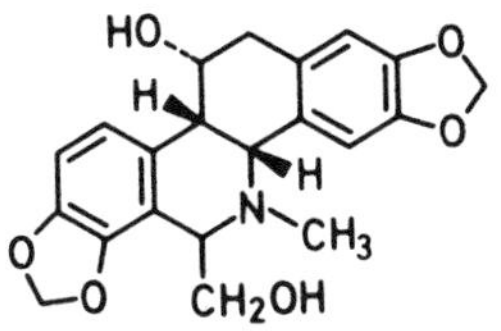

5/146, *Corynolamine* 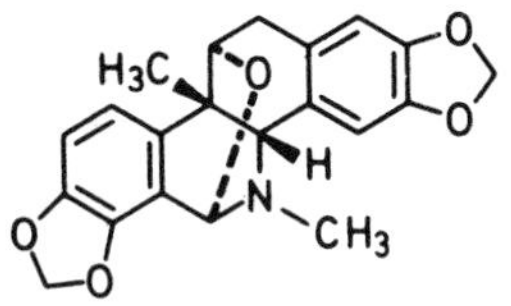   5/147, *Corynoloxine*

5.1.7.14 Spirobenzylisoquinoline Alkaloids [M 33], [S 49]

Fumaricine [5/148, from *Fumaria* sp. (FUM)] and ochotensine [5/149, from *Corydalis* and *Dicentra* spp. (FUM)] may be regarded as characteristic representatives of this group of alkaloids.

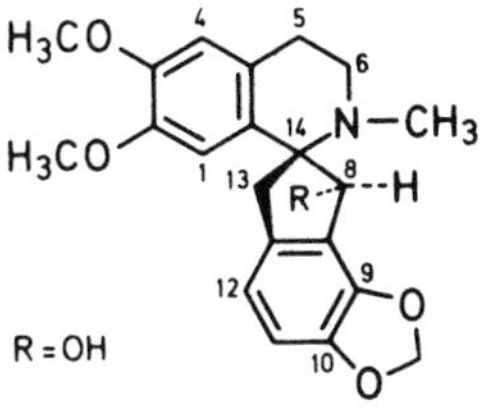

5/148, (−)-*Fumaricine* 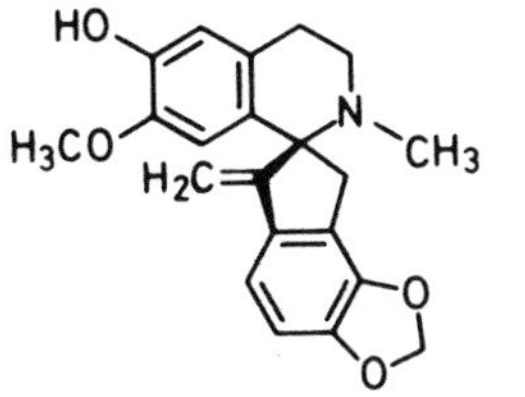   5/149, (+)-*Ochotensine*

5.1.7.15 Pavine and Isopavine Alkaloids

Argemonine [5/150, from *Argemone* sp. (PAP)] and amurensine [5/151, from *Papaver* sp. (PAP)] are examples of pavine and isopavine alkaloids, respectively.

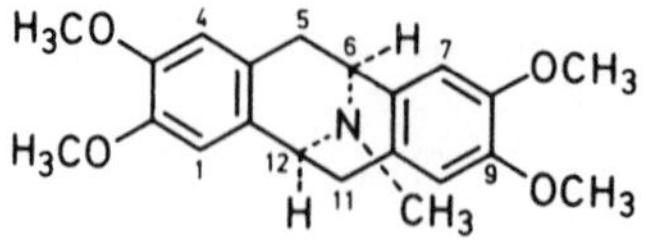

5/150, (−)-*Argemonine*

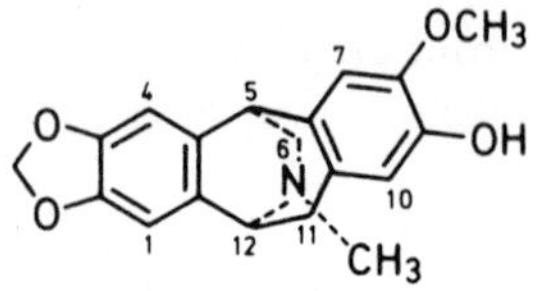

5/151, (−)-*Amurensine*

5.1.7.16 Morphine Alkaloids [B 6], [B 14], [B 33], [G 3], [H 7], [H 8], [S 18], [S 35] (See Section 6)

This well-known class of isoquinoline alkaloid may be divided into several subgroups: sinomenine [5/152, from *Menispermum* and *Sinomenium* spp. (MEN)] is a tetracyclic compound, while morphine [5/153, from *Papaver somniferum* (PAP)] has an extra ether ring. Hasubanan alkaloids [e.g., hasubanonine (5/154) from *Stephania japonica* (MEN)] are structurally related

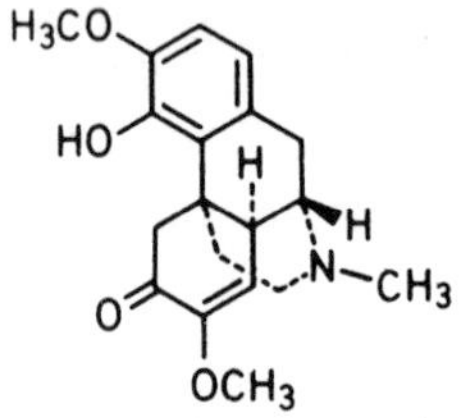

5/152, (−)-*Sinomenine [H 9]*

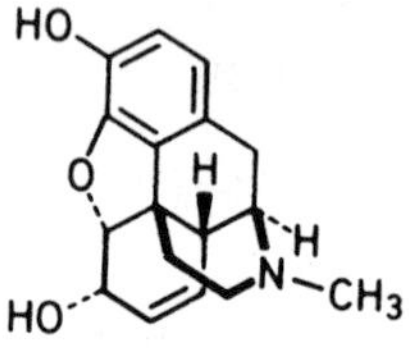

5/153, (−)-*Morphine*

to the true morphine alkaloids; likewise the homomorphine alkaloids, which have an enlarged ring as compared to the morphine bases, for example, androcymbine [5/155, from *Androcymbium* sp. (LIL)] and kreysiginine [5/156, from *Kreysigia* sp. (LIL)], also alkaloids of the acutumine type, such as acutumine [5/157, from *Menispermum* sp. (MEN)].

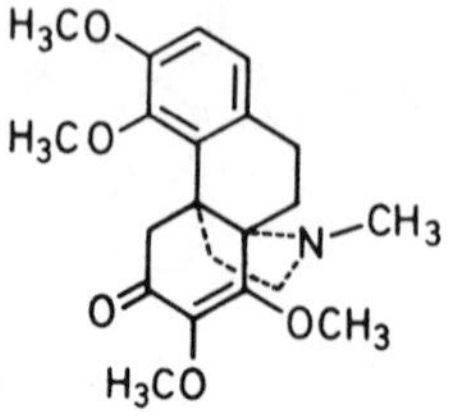

5/154, *Hasubanonine*

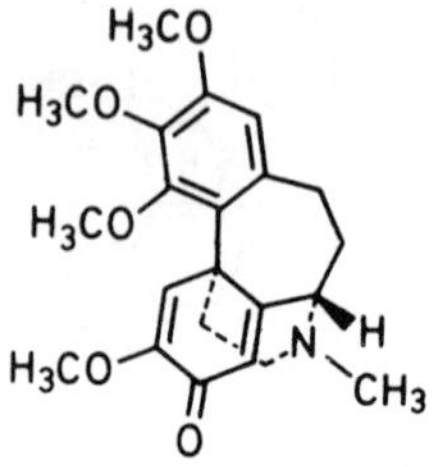

5/155, *Androcymbine*

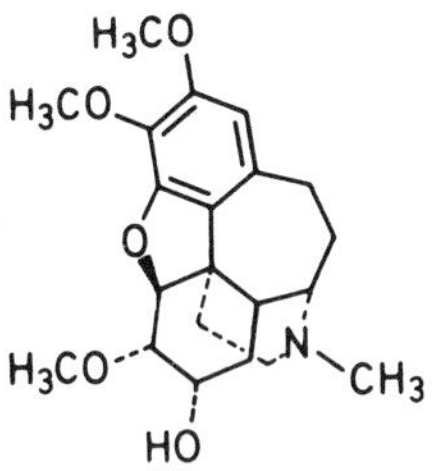

5/156, (+)-Kreysiginine

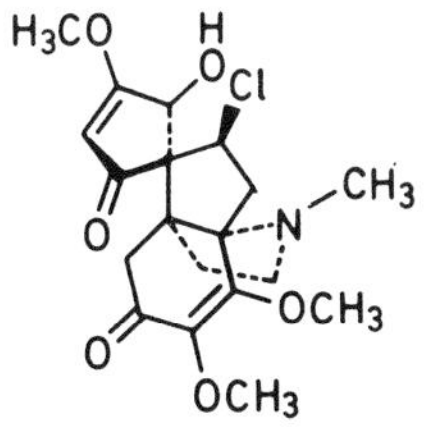

5/157, (−)-Acutumine

Morphine is very effective in eliminating pain, but it is dangerously addictive with repeated use. Codeine (O-methylmorphine) is a well-known specific cough remedy.

5.1.7.17 Amaryllidaceous Alkaloids [C 1], [F 7], [J 5], [W 1], [W 2], [W 3]

In contrast to the other isoquinoline alkaloids, the basis of classification for these alkaloids is their common occurrence in the plant family Amaryllidaceae. Six different skeletal types can be distinquished, which from their structural formulas are evidently all related to the fundamental type of lycorine:

Lycorine (5/158, widely distributed), provides an example of the basic type

Ambelline (5/159, from *Amaryllis, Ammocharis, Brunsvigia*, and *Buphane* spp.)

Hippeastrine (5/160, from *Amaryllis, Brunsvigia, Clivia*, and *Crinum* spp.)

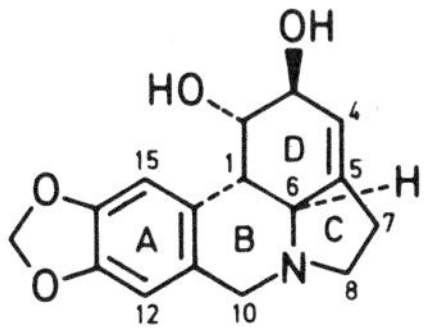

5/158, *Lycorine*

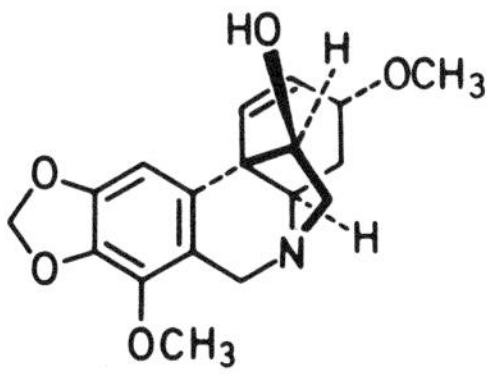

5/159, *Ambelline*

$\equiv$

R = CH₃

5/160, (+)-*Hippeastrine*

Tazettine (**5/161**, from various genera)

Galanthamine (**5/162**, from *Amaryllis, Cooperanthes*, and *Crinum* spp.)

Montanine (**5/163**, from *Haemanthus* sp.)

These fundamental types are found in many individual alkaloids, but the range of structural variation is small.

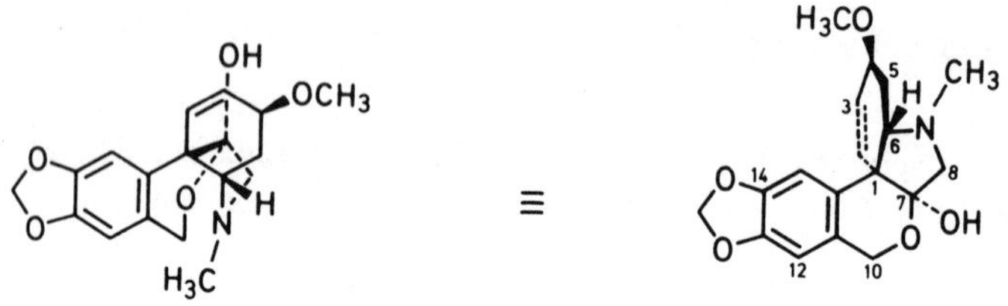

5/161, (+)-*Tazettine*

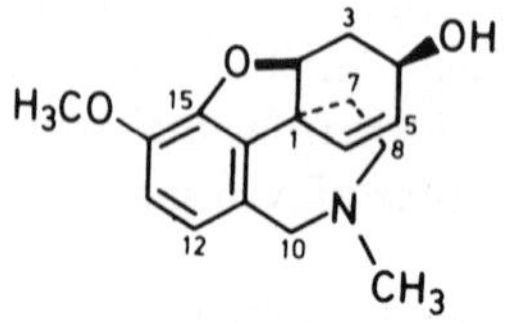

5/162, (−)-*Galanthamine* 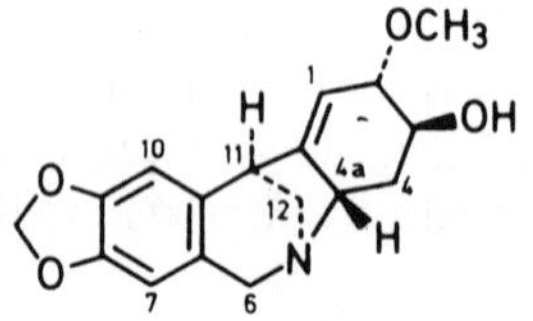   5/163, (−)-*Montanine*

5.1.7.18 Cherylline Type Alkaloids

So far the only member of this group is cherylline [**5/164**, from *Crinum* sp. (AMA)] , which has a clear structural relationship to the last-mentioned type.

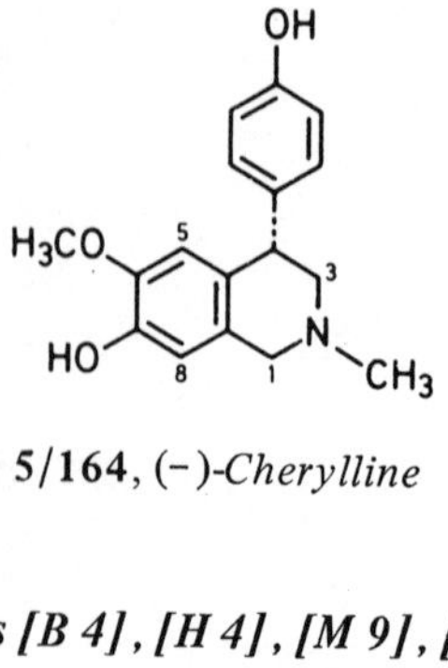

5/164, (−)-*Cherylline*

5.1.7.19 Erythrina Alkaloids [B 4], [H 4], [M 9], [M 10] (See Section 6)

The *Erythrina* alkaloids could be included with either the isoquinoline or the izidine alkaloids according to our method of classification; the decision to place them under the former heading rests on biogenetic considerations. Two skeletal types that are typical of this group are represented by erysodine [5/165, from *Erythrina* sp. (LEG); see Section, 9.3] and β-erythroidine (5/166, from *Erythrina* sp.).

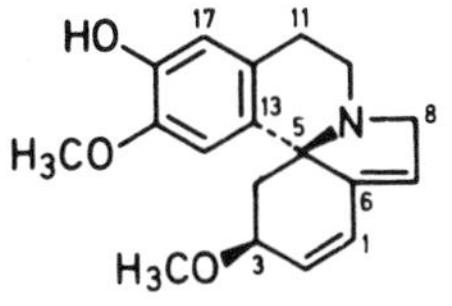

5/165, (+)-*Erysodine* 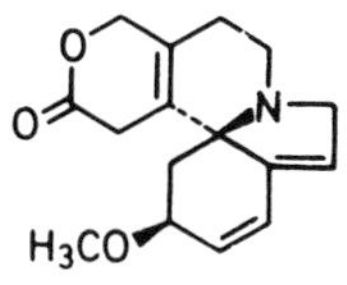   5/166, β-*Erythroidine*

A large number of naturally occurring examples have the same skeleton as 5/165 and differ only in the degree of dehydrogenation and in the pattern of substitution of the oxy functions. Homo derivatives have also been found in this group of isoquinoline bases, for example, schelhammerine [5/167, from *Schelhammera* sp. (LIL)]. Homo analogues of the lactone alkaloid of type 5/166 are likewise known. The cephalotaxus alkaloids, such as desoxyharringtonine [5/168, from *Cephalotaxus* sp. (CEP)], have a nucleus isomeric with that of the *Erythrina* alkaloids; they have antileucemic properties.

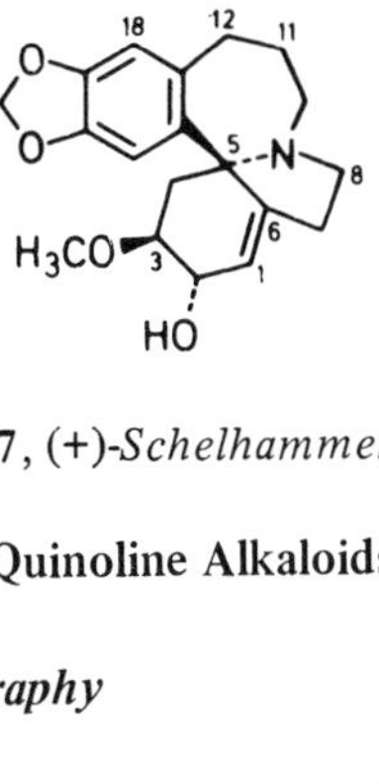

5/167, (+)-*Schelhammerine* 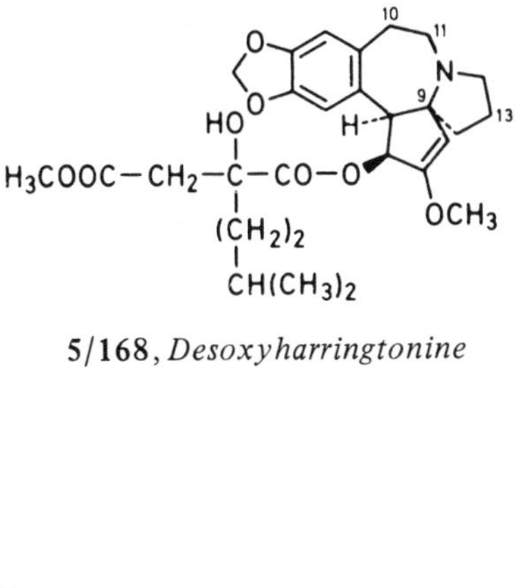   5/168, *Desoxyharringtonine*

5.1.8 Quinoline Alkaloids

Bibliography

General reviews: [O 2], [O 3], [O 4]

There are about 150 members of this class of alkaloid, in which the chromophores 5/169 to 5/174 can be distinguished.

5.1.8.1 *Quinolone-2 and Quinolone-4 Derivatives*

Apart from simple compounds such as echinopsine [5/175, from *Echinops* sp. (COM)] and *N*-methylquinolone-2 [from *Galipea* sp. (RUT)], various *O*-

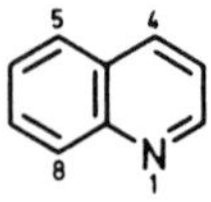

5/169, *Quinoline* 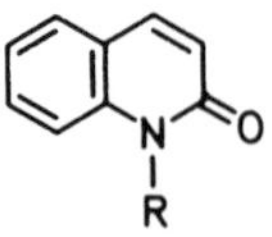   **5/170**, *Quinolone-2* 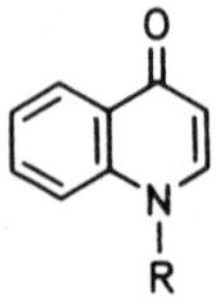   **5/171**, *Quinolone-4*

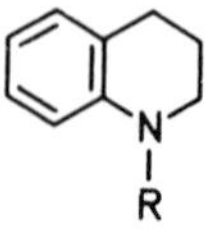

5/172, *Tetrahydroquinoline* 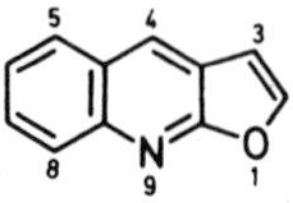   **5/173**, *Furoquinoline* 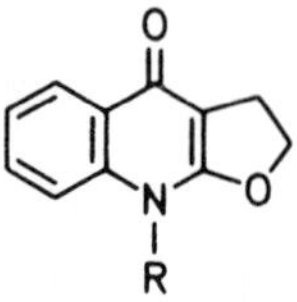   **5/174**, *Furoquinolone*

and *C*-alkylated derivatives also occur. There is considerable variation in the manner of alkyl substitution. In evocarpine [**5/176**, from *Evodia* sp. (RUT)] an unbranched C_{13}-chain is attached at C-2; in galipine (**5/177**, from *Galipea* sp.) a 2-phenylethyl residue occurs; in lunamarine [**5/178**, from *Lunasia* sp. (RUT)] there is a phenyl residue, and in ptelefolidine [**5/179**, from *Ptelea* sp. (RUT)] there is a prenyl group at C-3. Many structural variations of these types are known.

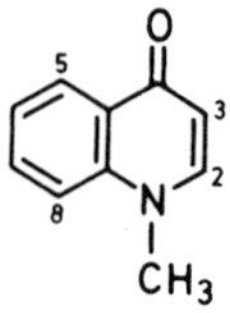

5/175, *Echinopsine*

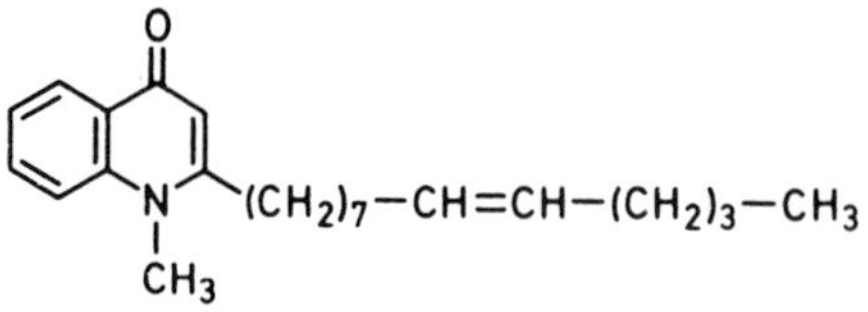

$(CH_2)_7-CH=CH-(CH_2)_3-CH_3$

5/176, *Evocarpine*

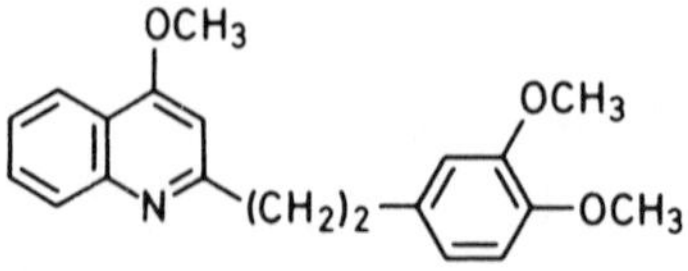

5/177, *Galipine*

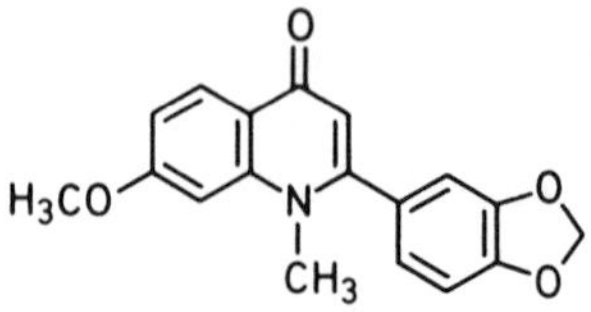

5/178, *Lunamarine*

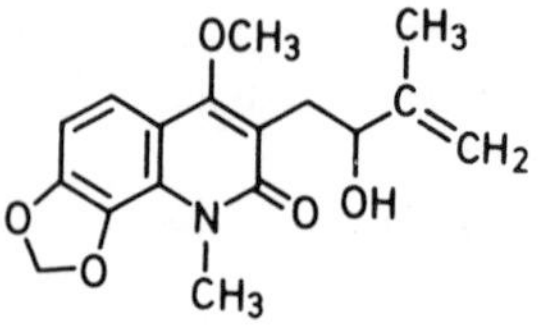

5/179, *Ptelefolidine*

5.1.8.2 Furoquinoline Alkaloids [S 11]

In addition to simple representatives like dictamnine (5/180, from various genera of FLI and RUT), there are also bases that are formed from isoprene derivatives of the type 5/179, such as lunacrine [5/181, from *Lunasia* sp. (RUT)]; acrophylline [5/182, from *Acronychia* sp. (RUT)] has an isoprene residue attached to nitrogen. Quaternary quinolinium salts have also been isolated from various plants.

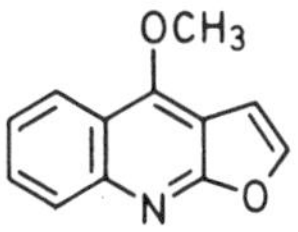

5/180, *Dictamnine* 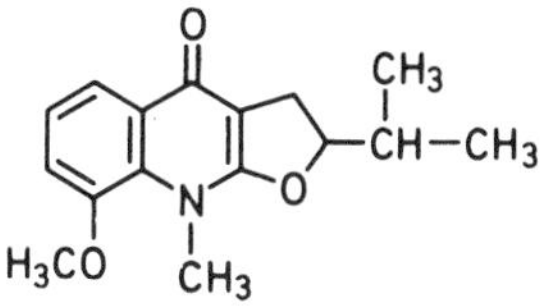   5/181, *Lunacrine* 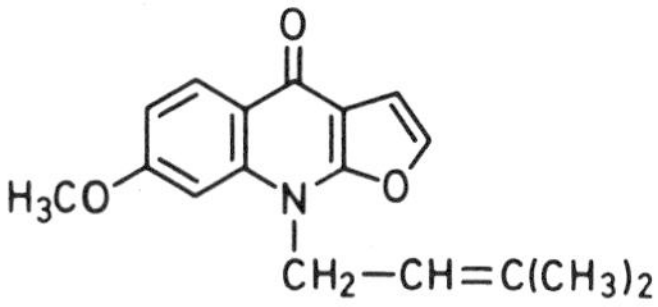   5/182, *Acrophylline*

5.1.8.3 Quinine Alkaloids (Cinchona Alkaloids) [S 8], [T 4], [U 1]

This group of alkaloids, which are some of the earliest known, can be divided into two skeletal types represented by quinine [5/183, from *Cinchona* and *Remijia* spp. (RUB)] and quinotoxine (5/184, from *Cinchona* sp.), which has a secoquinuclidine ring. The *Cinchona* alkaloids with an indole nucleus (see Section 5.1.2.3.4) are biogenetically related to these compounds. Quinine is important as an antimalarital and febrifuge.

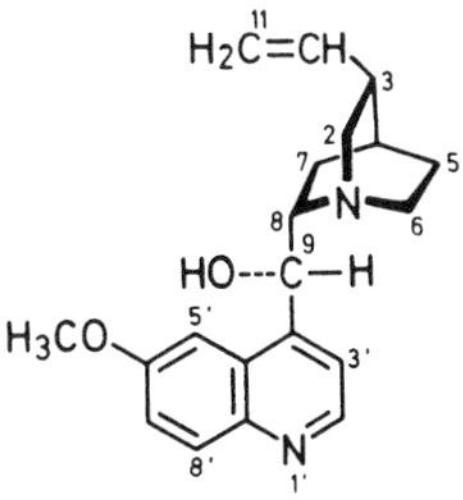

5/183, (−)-*Quinine* 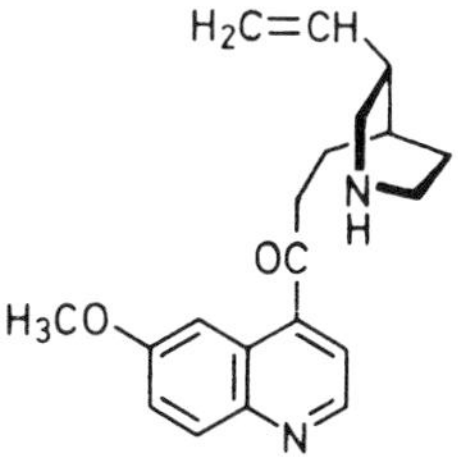   5/184, (+)-*Quinotoxine*

5.1.8.4 Melodinus *Alkaloids with a Quinoline Nucleus*

Although this small group of alkaloids belongs biogenetically to the indole alkaloids of the aspidospermine type (see Section 5.1.2.2.2), they are included here in accord with our classification [e.g., meloscine (5/185) from *Melodinus* sp. (APO)].

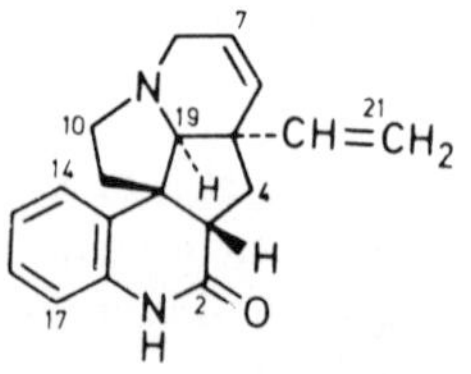

5/185, (+)-Meloscine

5.1.8.5 Camptotheca *Bases*

Camptothecine [5/186, from *Camptotheca* (NYS) and *Mappia* (OLA) spp.] is the main representative of this structural type, and like 5/185 is also related biogenetically to the indole alkaloids; these bases have antileucemic properties.

Bisalkaloids with a tetrahydroquinoline chromophore have also been isolated from *Calycanthus* species (see Section 10.4).

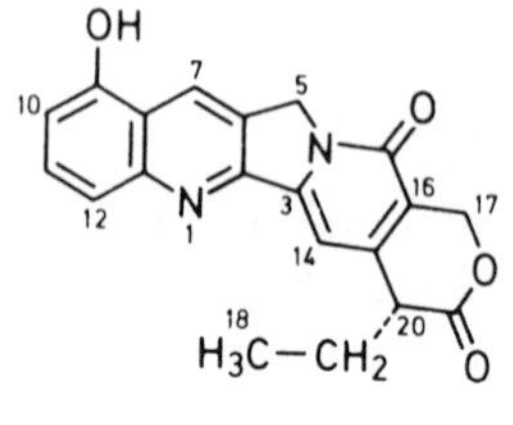

5/186, (+)-*Camptothecine*

5.1.9 Acridine Alkaloids

Bibliography

General reviews: [P 2]

This group contains about 40 bases. In addition to simple acridine derivatives, such as evoxanthine [5/187, from *Evodia* and *Teclea* spp. (RUT)] , examples are also known with prenyl substituents, such as atalaphylline [5/188, from *Atalantia* sp. (RUT)] . Tetra- and pentacyclic derivatives with extra ether rings also occur.

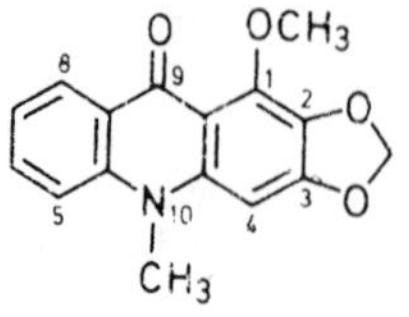

5/187, *Evoxanthine*

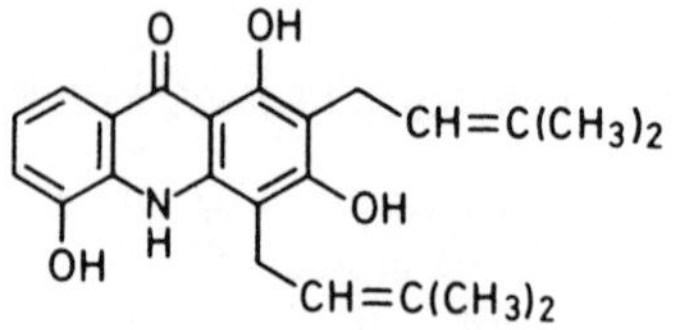

5/188, *Atalaphylline*

5.1.10 Quinazoline Alkaloids

Bibliography

General reviews: [O 5] , [O 6]

So far about 40 alkaloids with a quinazoline ring **5/189** have been isolated. Apart from simple derivatives such as glomerine (**5/190**, from *Glomeris* sp. of diplopod; animal defense substance), there are also derivatives that have one or two additional ring systems. Examples are provided by peganine (vasicine) (**5/191**) from *Adhatoda* (ACA), *Linaria* (SCR), and *Peganum* (ZYG) spp. and vasicolinone (**4/4**, from *Adhatoda vasica*). Febrifugine [**5/192**, from *Dichroa* sp. (HDR)] exhibits a further structural variation; moreover alkaloids of the evodiamine type which have been included among the indole alkaloids could equally well be classified as quinazoline alkaloids (see Section 5.1.2.8).

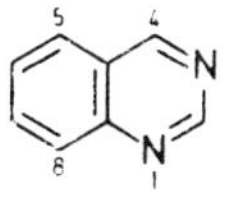

5/189, *Quinazoline* 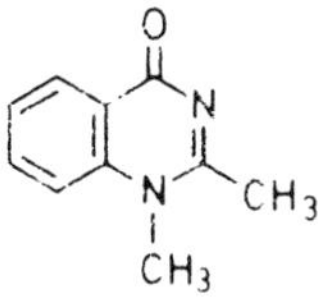   **5/190**, *Glomerine* 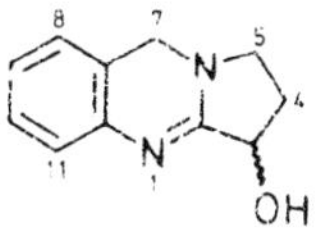   **5/191**, *Peganine*

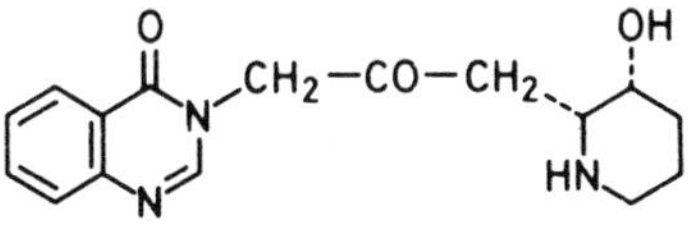

5/192, *Febrifugine*

5.1.11 Izidine Alkaloids

This section comprises about 500 alkaloids with nuclei in which the nitrogen atom forms part of two or three different rings. It includes alkaloids of the pyrrolizidine (**5/193**), indolizidine (**5/194**), and quinolizidine (**5/195**) types (see Section 11.2).

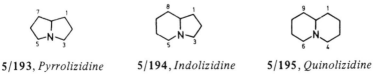

5/193, *Pyrrolizidine*   **5/194**, *Indolizidine*   **5/195**, *Quinolizidine*

5.1.11.1 Pyrrolizidine Alkaloids (Senecio Alkaloids)

Bibliography

General reviews: [F 8] , [L 5] , [L 6] , [W 8] , [W 9] , [W 10]

Four main alkaloid types can be distinguished under this heading. The alkaloids that have the necine skeleton as essential nucleus, such as retronecanol [5/196, from *Crotalaria* sp. (LEG)] , are comparatively rare in relation to the other groups. The ester alkaloids in which one hydroxyl group is esterified, such as heliotrine [5/197, from *Heliotropium* sp. (BOR)] , form a second group. Lasiocarpine [5/198, from *Heliotropium*, *Lappula*, and *Symphytum* spp. (BOR)] forms an example of the diester type. Monocrotaline [5/199, from *Crotalaria* sp. (LEG)] , which consists of necine and a dicarboxylic acid, constitutes another diester type. Structural variants include compounds in which C-9 is in the form of a carboxy group; also quaternary bases, *N*-oxides, and compounds with the pyrrolizidine ring opened, such as clivorine [5/200, from *Ligularia* sp. (COM)] . The large number of representatives of this group arises mainly through variation in the carboxylic acid component.

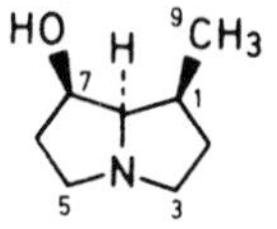

5/196, (–)-*Retronecanol*

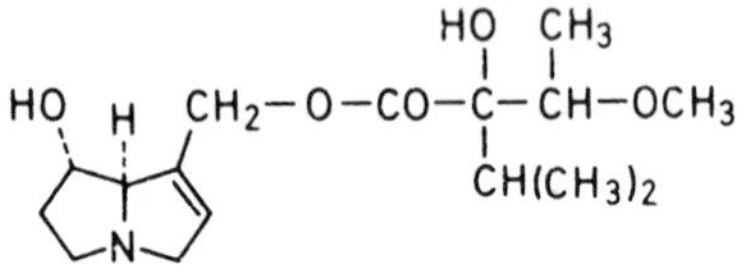

5/197, *Heliotrine*

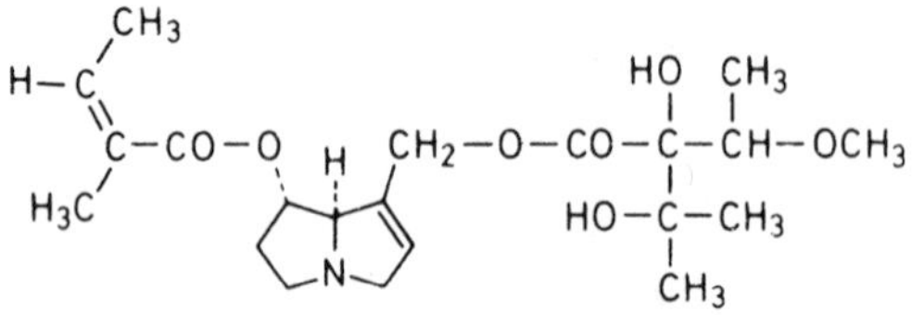

5/198, *Lasiocarpine*

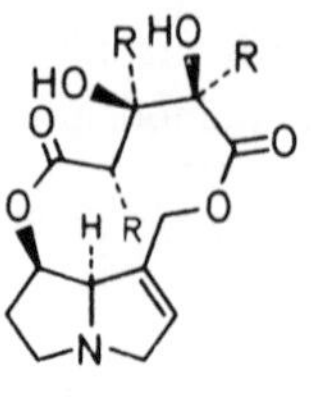

R = CH₃

5/199, *Monocrotaline*

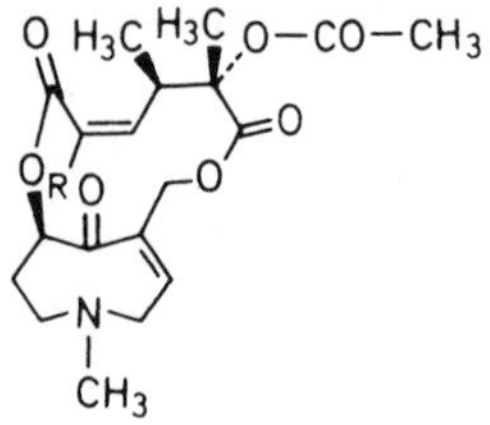

R = CH=CH₂

5/200, *Clivorine*

5.1.11.2 Indolizidine Alkaloids

This group includes alkaloids of the following types:

Slaframine (5/201, from *Rhizoctonia* sp., fungus)

Elaeokanine E [5/202, from *Elaeocarpus* sp. (ELE)]

Elaeocarpine (5/203, from *Elaeocarpus* sp.) [J 8]

Crepidamine [5/204, from *Dendrobium* sp. (ORC)]

Serratinine [5/205, from *Lycopodium* sp. (LYC)]

Securitinine [5/206, from *Securinega* sp. (EUP)]

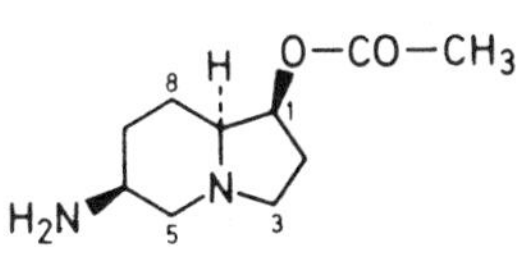

5/201, *Slaframine*

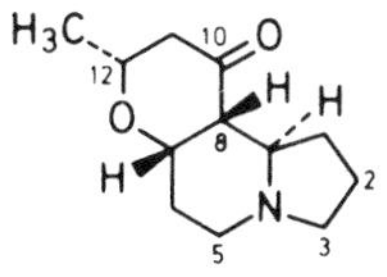

5/202, *Elaeokanine E*

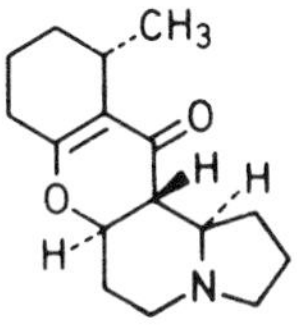

5/203, *Elaeocarpine*

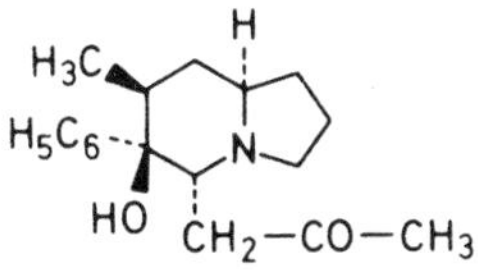

5/204, *Crepidamine*

The *Securinega* alkaloids [S 53] to which 5/206 belongs also include a pyrrolizidine alkaloid, norsecurinine (5/207). Certain alkaloids are classified in this group that are biogenetically related to the isoquinoline alkaloids, but which have an indolizidine, quinolizidine, or related nucleus; for example, tylophorine [5/208, from *Cynanchum*, *Tylophora*, and *Vincetoxicum* spp. (ASC) and *Ficus* sp. (MOR); see section 9.2), cryptopleurine [5/209, from *Cryptocarya* sp. (LAU)] and protostephanine [5/210, from *Stephania* sp. (MEN)] [G 4]. It is possible that 5/204 is an artefact (see Section 10.1).

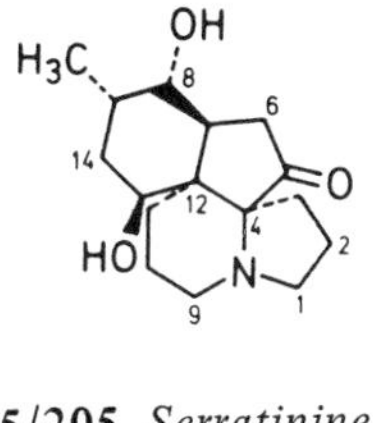

5/205, *Serratinine*

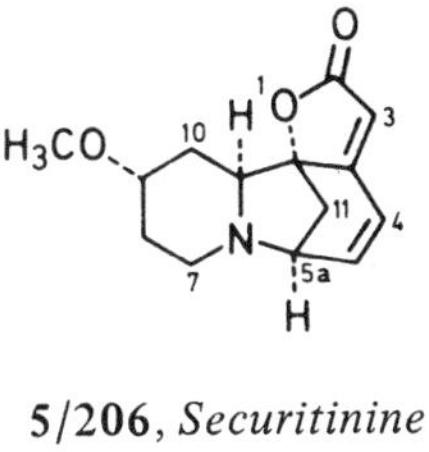

5/206, *Securitinine*

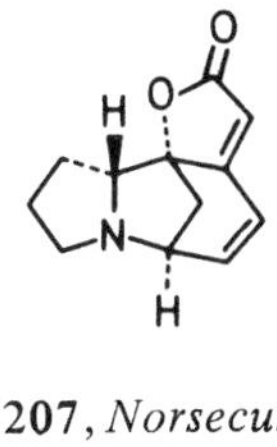

5/207, *Norsecurinine*

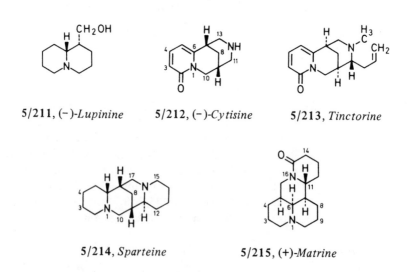

5/208, (S)-(−)-Tylophorine   5/209, Cryptopleurine   5/210, Protostephanine

5.1.11.3 Quinolizidine Alkaloids

5.1.11.3.1 Lupin Alkaloids [B 5], [G 6], [L 3], [L 4] This category includes the bicyclic lupinine (5/211) from *Anabasis* (CHE) and *Lupinus* (LEG) spp., the tricyclic cytisine (5/212, very toxic, widely distributed in the family LEG), tinctorine [5/213, from *Genista* sp. (LEG)], and the tetracyclic sparteine [5/214, widely distributed in LEG, also in *Chelidonium* sp. (PAP) and in LEO]. Matrine [5/215, from *Sophora* sp. (LEG)] is also included in this group. There are many naturally occurring derivatives of these various types (see Section 10).

5/211, (−)-Lupinine   5/212, (−)-Cytisine   5/213, Tinctorine

5/214, Sparteine   5/215, (+)-Matrine

5.1.11.3.2 Nuphar Alkaloids [W 7] Most *Nuphar* alkaloids contain a quinolizidine nucleus. Desoxynupharidine [5/216, from *Nuphar* sp. (NYM)] can be regarded as the fundamental type.

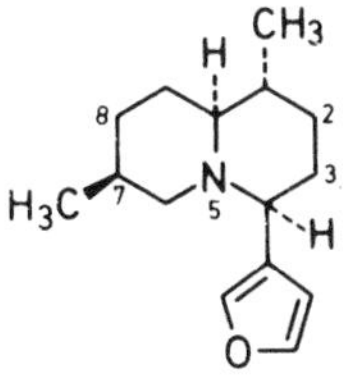

5/216, (−)-*Desoxynupharidine*

5.1.11.3.3 Lythraceous Alkaloids Two fundamental types have so far been isolated, of which lythrine [5/217, from *Decodon* and *Heimia* spp. (LYT)] and lythrancine III [5/218, from *Lythrum* sp. (LYT)] provide examples.

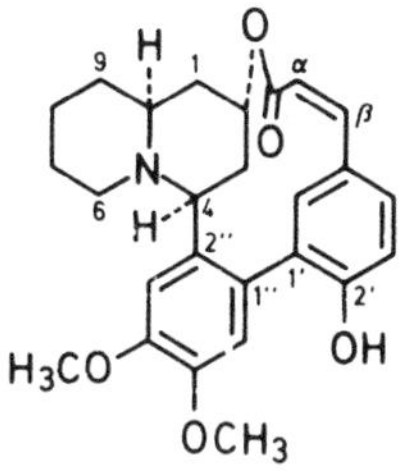

5/217, *Lythrine* 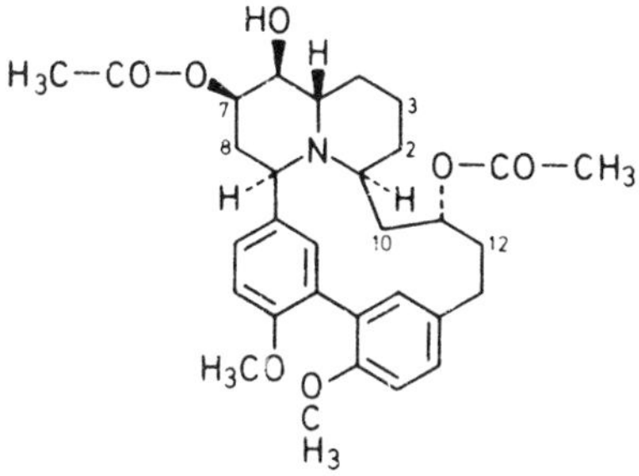   5/218, *Lythrancine III*

5.1.11.3.4 Lycopodium Alkaloids [A 3], [M 16], [M 17], [M 18], [M 41], [W 11] The alkaloids isolated from *Lycopodium* sp. (LYC) include various structural types, but most of them have a quinolizidine group; for the serratinine type, see Section 5.1.11.2. Examples of various individual types are shown below. (See also Sections 9.3 and 11.2.)

Lycopodine
(Lycopodine type)

Lyconnotine
(Lyconnotine type)

Annotine
(Annotine type)

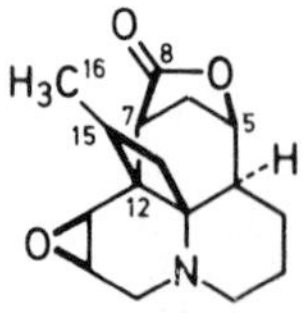

Annotinine
(Annotinine type)

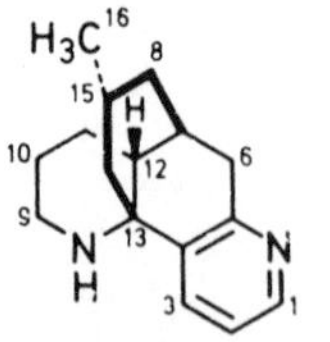

(−)-Lycodine
(Lycodine type)

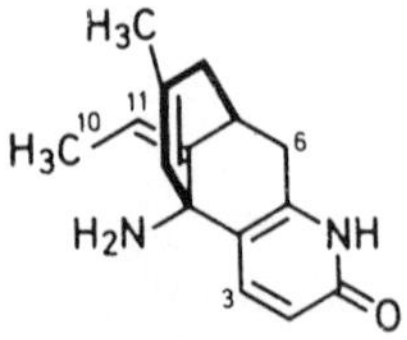

Selagine
(Selagine type)

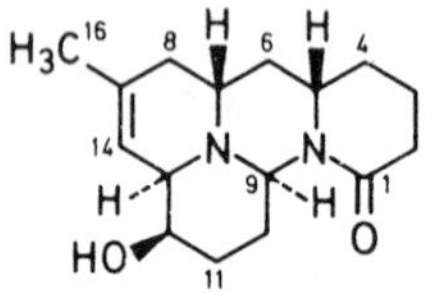

Carolinianine
(Cernuine type)

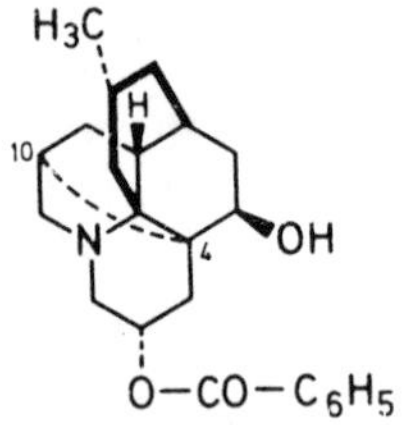

(−)-Alopecurine
(Inundatine type)

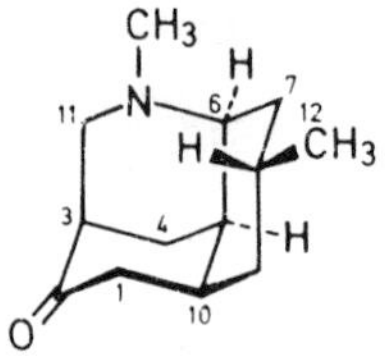

(+)-Luciduline
(Luciduline type)

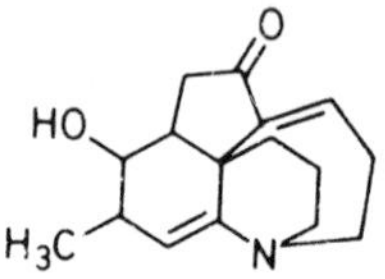

Alolycopine

5.2 ALKALOIDS WITH AN EXOCYCLIC NITROGEN

Four categories of alkaloids are included under this heading.

1 The *Erythrophleum* bases [e.g., cassaine (5/219) from *Erythrophleum* sp. (LEG)] , which have an aminoethanol grouping [D 2] , [M 11] .

2 The phenylalkylamines [F 8] , [K 14] , [R 2] , [R 3] , such as ephedrine (5/220), from plants of the genus *Ephedra* (EPH). *Ephedra* extracts have been used for centuries in Chinese medicine for throat complaints, and (−)-ephedrine is an oral sympathomimetic drug. This group also includes a range of compounds that may be considered as being formed from the corresponding isoquinoline alkaloids by a Hofmann degradation [e.g., uvariopsamine (5/221) from *Uvariopsis* sp. (ANN)] .

3 Capsaicine (5/222), from *Piper* (PIP) and *Capsicum* (SOL) spp., is an example of the benzylamine type of alkaloids.

4   Alkaloids of the colchicine type have an exocyclic nitrogen and are therefore
included here, although they are more closely related biogenetically to the
isoquinoline alkaloids, such as the very toxic base colchicine (see 2/2) from
Colchicum sp. (LIL); [C 5], [W 4], [W 5], [W 6].

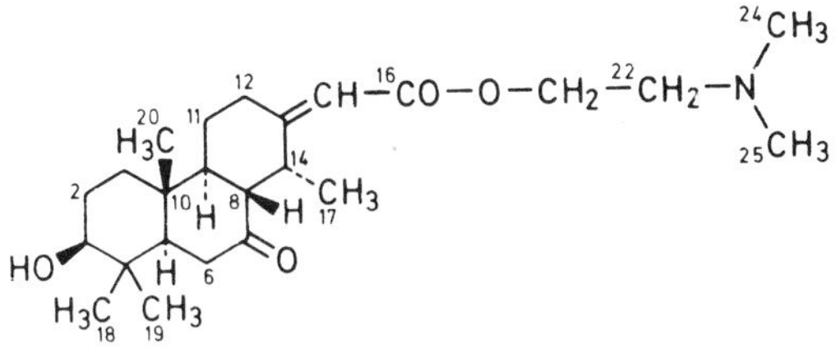

5/219, (−)-*Cassaine*

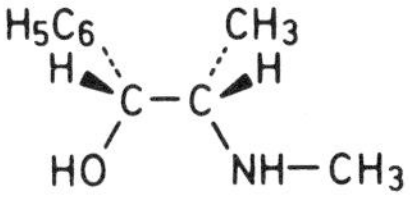

5/220, (−)-*Ephedrine*

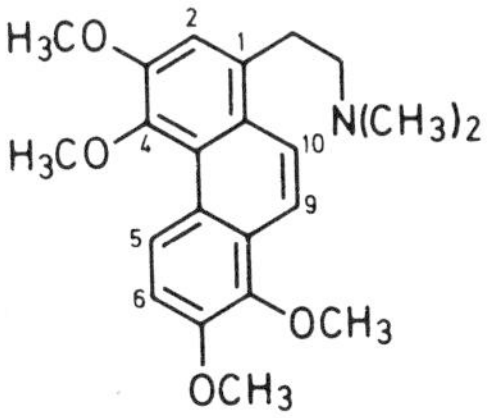

5/221, *Uvariopsamine*

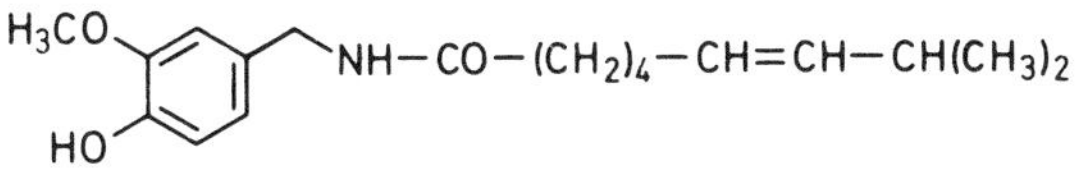

5/222, *Capsaicine*

5.3   PUTRESCINE, SPERMIDINE, AND SPERMINE ALKALOIDS [B 23], [H 20]

As already mentioned, these three bases belong to the biogenic amines, but
their derivatives (mostly containing fatty acid or cinnamic acid residues) are
considered to be alkaloids. Typical examples are paucine [5/223, from *Penta-
clethra* sp. (LEG)], inandenine-12-one [5/224, from *Oncinotis* sp. (APO)],
codonocarpine [5/225, from *Codonocarpus* sp. (GYR)], and chaenorhine
[5/226, from *Chaenohinum* sp. (SCR)]. (See Sections 9.1 and 11.3).

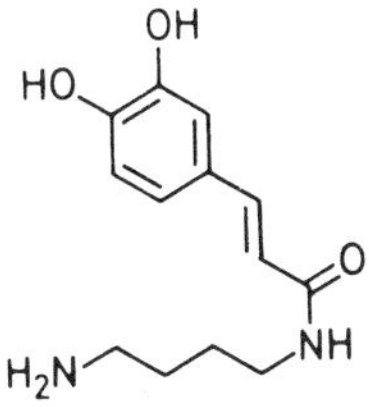

5/223, *Paucine*

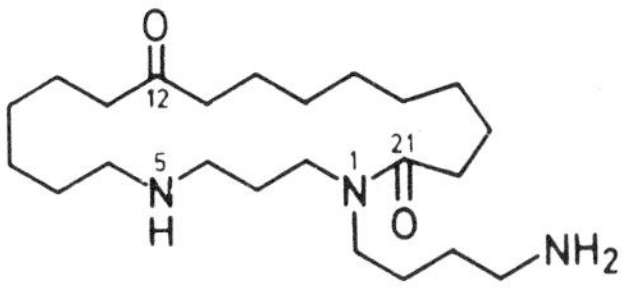

5/224, *Inandenine-12-one*

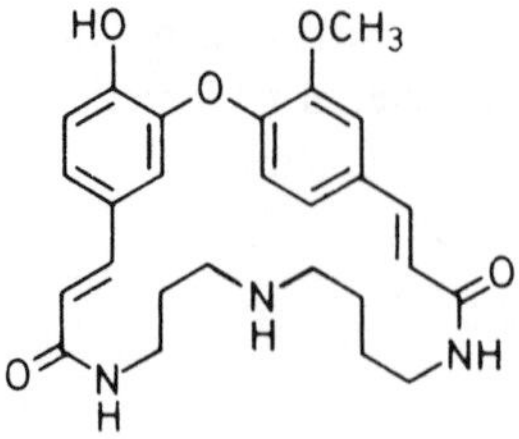

5/225, *Codonocarpine*

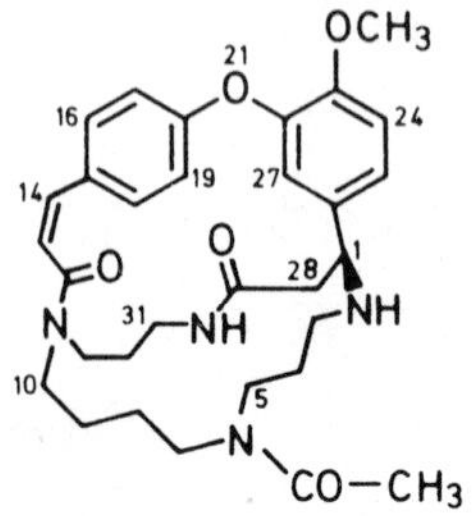

5/226, (+)-*Chaenorhine*

5.4 PEPTIDE ALKALOIDS [T 19], [W 15]

The ergot alkaloids, a subgroup of the peptide bases, have already been dealt with in connection with indole alkaloids (Section 5.1.2.7). This section comprises alkaloids of the Rhamnaceae, which include two main structural types: integerrine [5/227, from *Ceanothus* sp. (RHA)], consisting of the amino acids tryptophan, *N,N*-dimethylvaline, and phenylserine, together with *p*-hydroxystyrylamine as the amino component; and mucronine-A [5/228, from *Zizyphus* sp. (RHA).

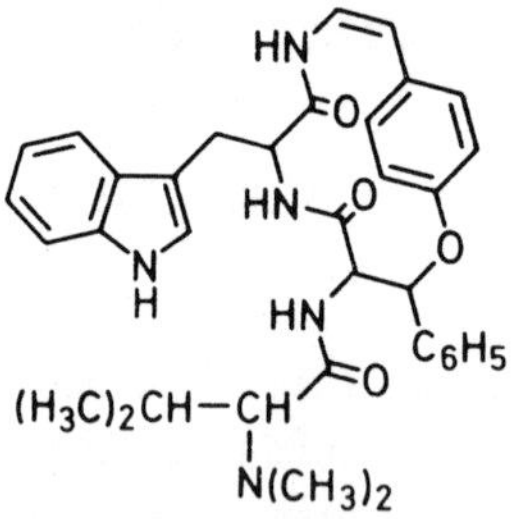

5/227, *Integerrine*

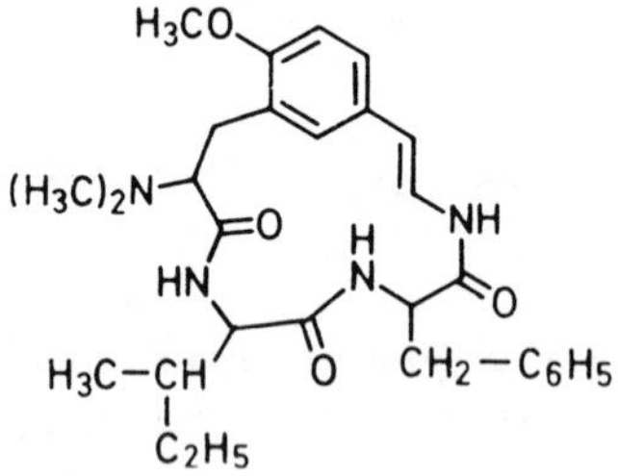

5/228, *Mucronine*

5.5 DITERPENE ALKALOIDS [K 3], [P 3], [P 4], [P 5], [P 9], [P 12], [S 9], [S 10], [W 12], [Y 2]

Monoterpene and sesquiterpene alkaloids have already been dealt with elsewhere (see under pyrrolidine, piperidine, and *Nuphar* alkaloids). Two large groups of diterpene alkaloids can be distinguished: those with a C_{20} skeleton and those with a C_{19}. They fall into four different skeletal types: the veatchine type [5/229; e.g., veatchine (5/230) from *Garrya* sp. (GAR)], the atisine type [5/231; e.g., atisine (5/232) from *Aconitum* sp., etc. (RAN)], the lycoctonine type [5/233, e.g., aconitine[3] (5/234) from *Aconitum* sp. (RAN)], and the heteratisine type [5/235, e.g., heteratisine (5/236) from *Acontium* sp.]. The large number of naturally occurring diterpene alkaloids (ca. 150 examples) differ from the above-mentioned fundamental types mainly in their substitution patterns.

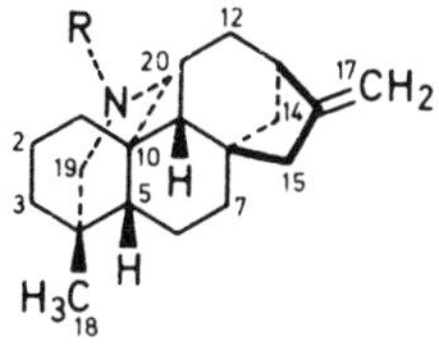

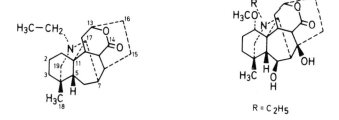

5/229, *Veatchine skeleton* 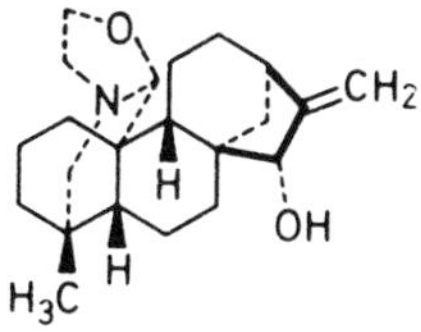   5/230, *Veatchine* 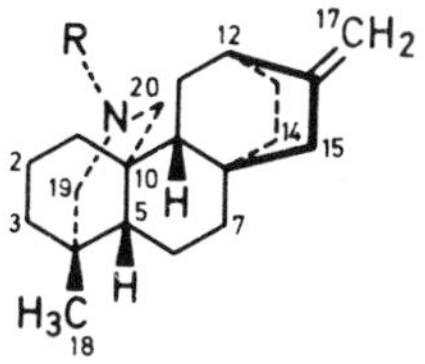   5/231, *Atisine skeleton*

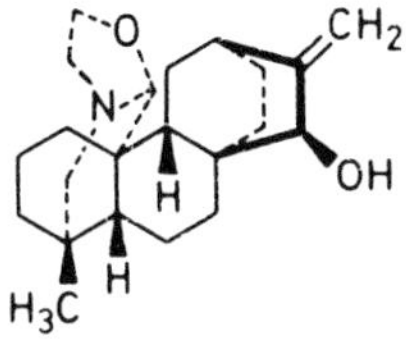

5/232, *Atisine* 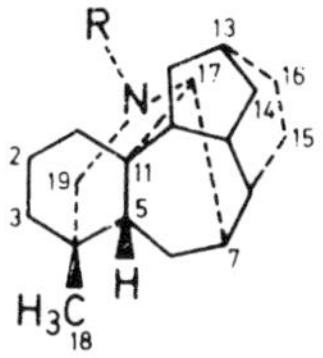   5/233, *Lycoctonine skeleton*   5/234, *Aconitine*

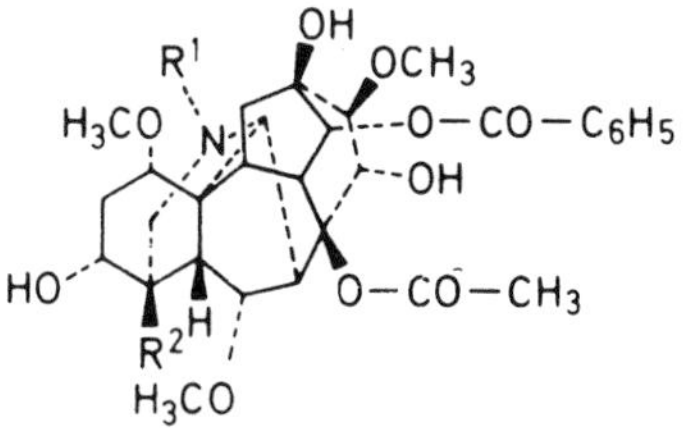

$R^1 = C_2H_5$
$R^2 = CH_2-OCH_3$

5/235, *Heteratisine skeleton*   5/236, *Heteratisine*

$R = C_2H_5$

---

[3] Aconitine and related alkaloids are extremely toxic compounds which act on the central nervous system.

5.6 STEROIDAL ALKALOIDS

Bibliography

General: [H 18], [M 8], [S 27]

Abnormal steroidal alkaloids: [B 9]

Alkaloids of the Apocynaceae: [C 8]; *Buxus*: [C 8], [T 18]; *Holarrhena*: [J 3]; *Solanum*: [P 7], [P 8], [S 28]; *Taxus*: [L 8]; *Veratrum*: [J 4], [K 4], [M 30], [N 2], [P 7], [T 18]

Salamander alkaloids: [H 10]

Steroidal alkaloids have an intact or modified steroid skeleton with the nitrogen forming an integral part, either incorporated into a ring or attached to the nucleus by a side chain. These bases are all included in the same class of steroidal alkaloids, irrespective of the nitrogen function.

Steroidal alkaloids occur in both animals and plants. Samandarine is an example of an alkaloid of animal origin (see 2/7). Various skeletal types may be distinguished on the basis of the steroid nucleus and the manner in which the nitrogen is incorporated; the following are examples:

Funtumine [5/237, from *Funtumia* sp. (APO)]

Paravallarine [5/238, from *Paravallaris* sp. (APO)]

Terminaline [5/239, from *Pachysandra* sp. (BUX)]

Holarrhimine [5/240, from *Holarrhena* sp. (APO)]

Conessine (5/241, from *Holarrhena* sp., see Section 9.2)

Veralkamine [5/242, from *Veratrum* sp. (LIL)]

Solasodine [5/243, from *Solanum* sp. (SOL)]

Rubijervine (5/244, from *Veratrum* sp.)

Solanocapsine (5/245, from *Solanum* sp.)

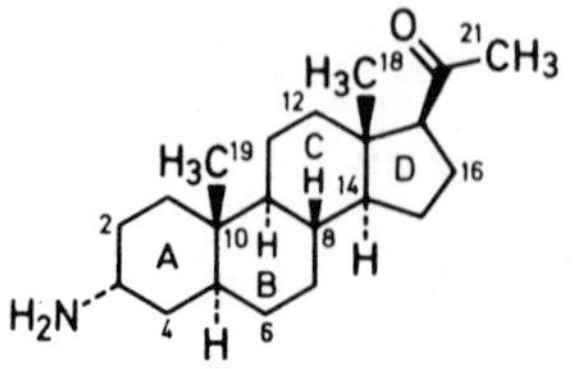

5/237, *Funtumine* 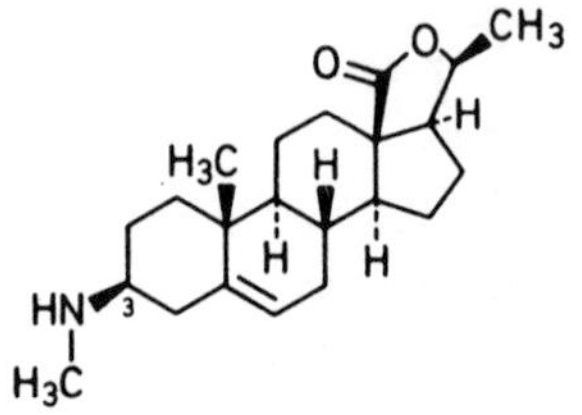   5/238, *Paravallarine*

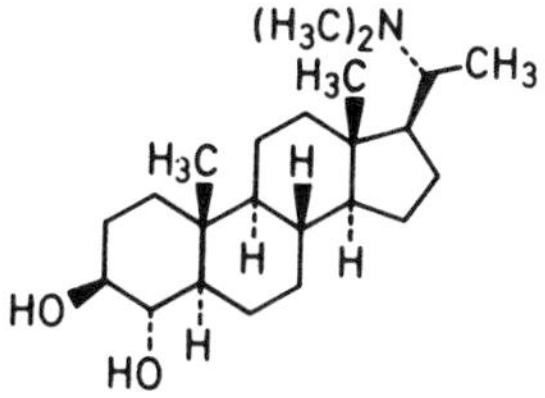

5/239, *Terminaline*

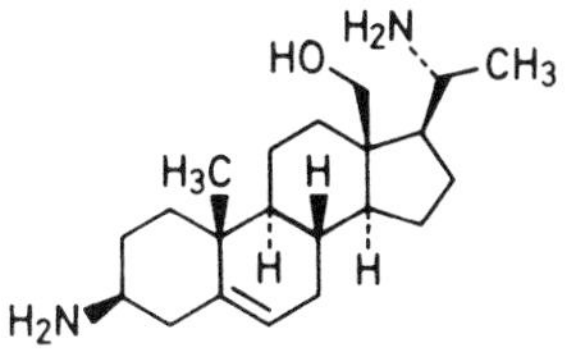

5/240, *Holarrhimine*

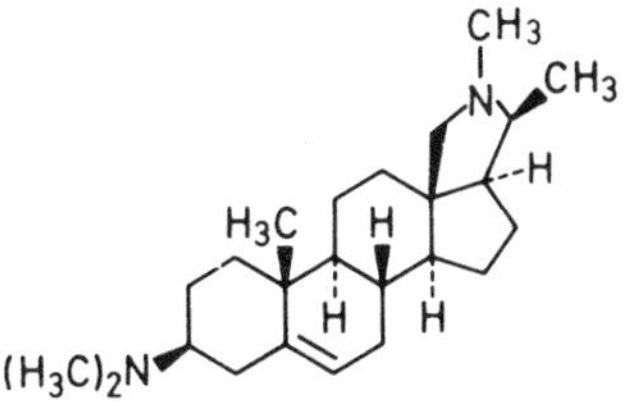

5/241, (−)-*Conessine*

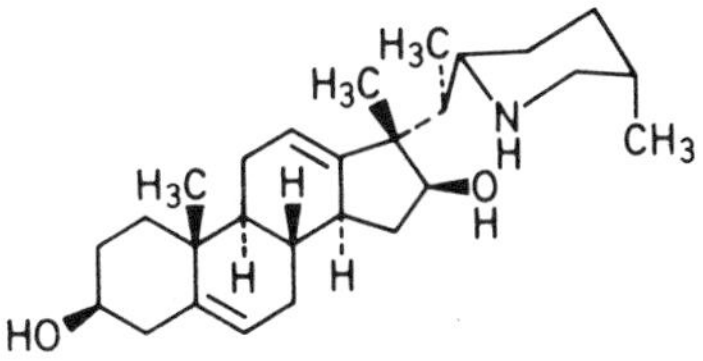

5/242, (−)-*Veralkamine*

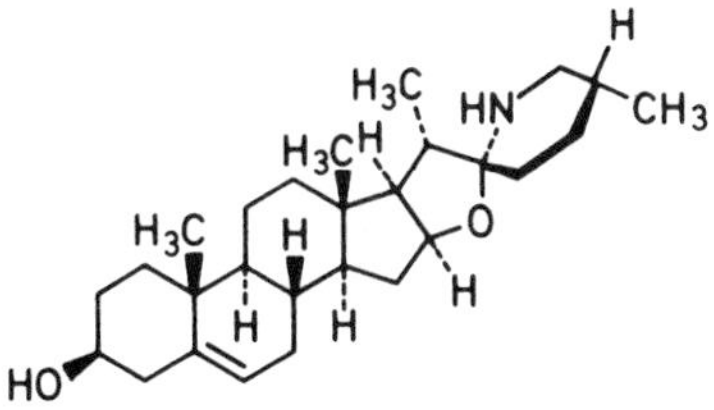

5/243, (−)-*Solasodine*

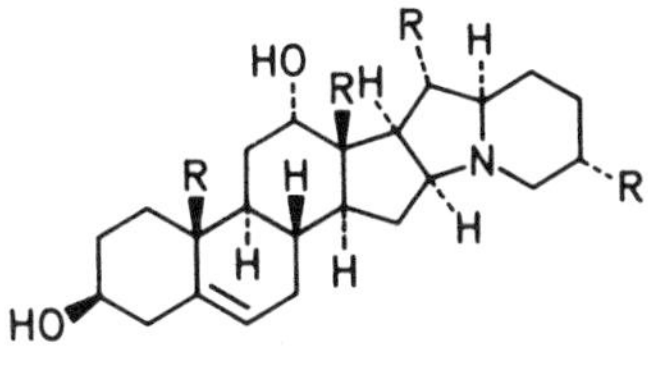

R = CH₃

5/244, *Rubijervine*

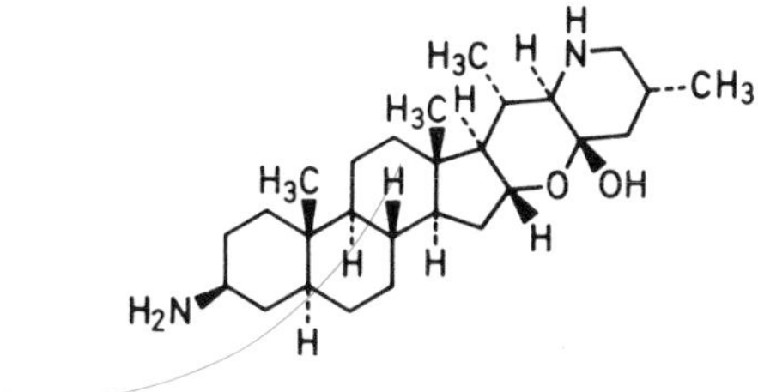

5/245, (+)-*Solanocapsine*

Jervine [5/246, from *Veratrum* and *Zigadenus* spp. (LIL)] and verticine [5/247, from *Fritillaria* sp. (LIL)] have a C-*nor*-D-homosteroidal skeleton. The *Buxus* alkaloid buxenine-G [5/248, from *Buxus* sp. (BUX)] has ring B enlarged through C-19, and cyclobuxidine-F (5/249, from *Buxus* sp.) has an additional cyclopropane ring.

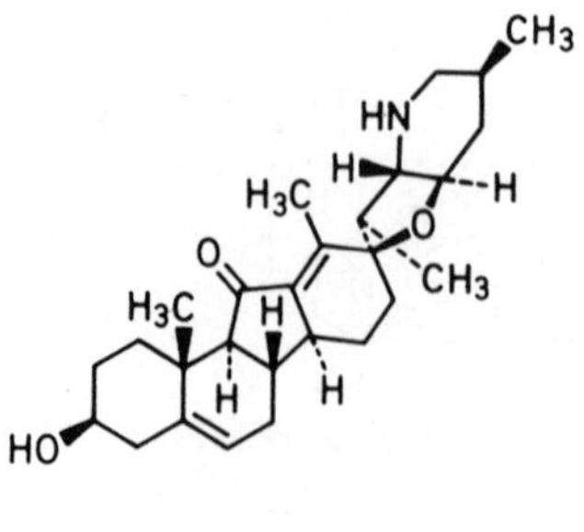

5/246, (−)-Jervine

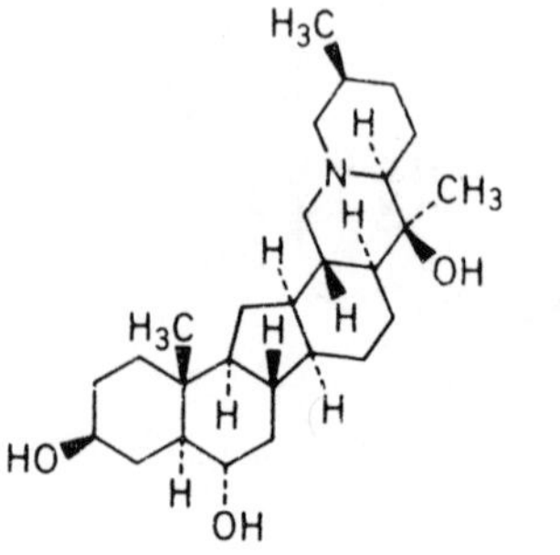

5/247, Verticine

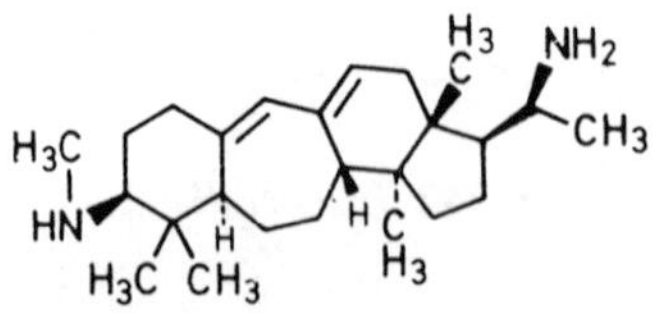

5/248, Buxenine G

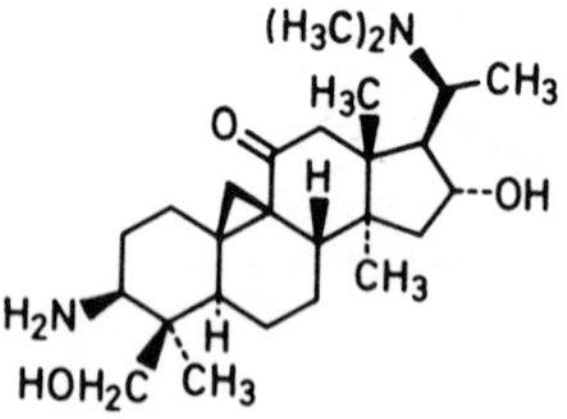

5/249, Cyclobuxidine F

The examples given throughout this section do not comprise by any means an exhaustive list; they merely represent a selection of the more important skeletal types.

Aspects of
Alkaloid Biogenesis

The term biogenesis refers to the manner in which organic substances are synthesized, altered, or degraded by plant or animal organisms. Alkaloids are of particular interest as far as biogenesis is concerned.

Although the same alkaloid may occur in several different organisms, the biochemical reactions that lead to its formation are not necessarily the same. A well-known example is nicotinic acid (6/1). One method of formation starts from tryptophan, which is degraded in a multistep process involving kynurenine (6/2) as one stage; alternatively nicotinic acid can be formed from glycerol and aspartic acid. This example is particularly striking because evidence has shown that certain organisms can synthesize nicotinic acid by either of these routes, and it underlines the need for caution in making generalizations from evidence of the mechanism of formation. A biogenetic pathway which has been established experimentally for a particular alkaloid strictly applies only for that compound in the organism used in the study; generalizations involving other compounds and other organisms can only be made with reservations. By rights one should thus take into account the results for each alkaloid and each organism that have so far been investigated, but such a compilation would lie outside the scope of the present book. We shall therefore adopt the plan of utilizing by way of illustration studies that have been carried out on a single class of bases. Some typical isoquinoline alkaloids have been selected as an example, and to give a better overall picture, various aspects of the biogenesis of 1-benzylisoquinoline alkaloids and of certain bases derived from them will be discussed.

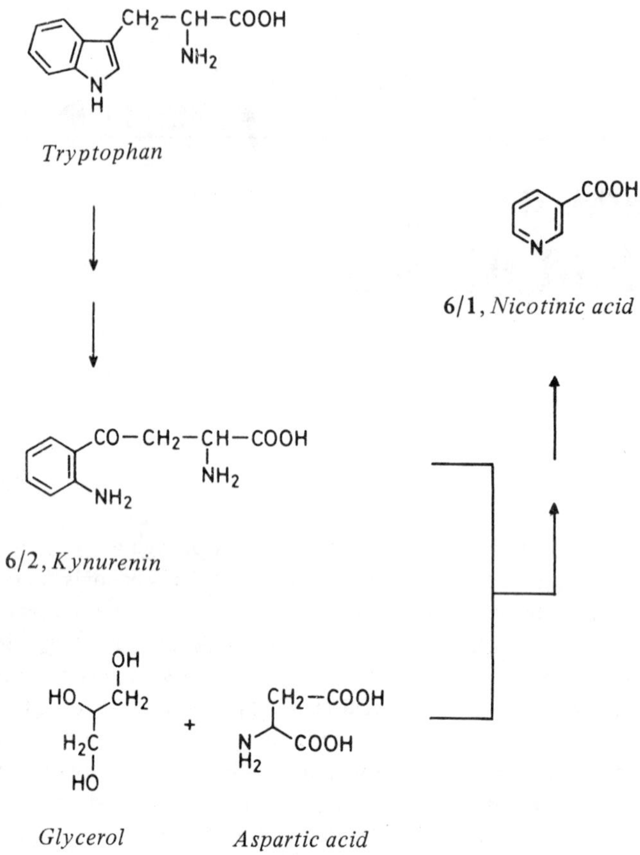

CH$_2$—CH—COOH
                    |
                   NH$_2$

Tryptophan

.COOH

6/1, *Nicotinic acid*

CO—CH$_2$—CH—COOH
                      |
                     NH$_2$

NH$_2$

6/2, *Kynurenin*

OH
|
HO   CH$_2$

H$_2$C
|
HO

+

CH$_2$—COOH
|
N   COOH
H$_2$

Glycerol   *Aspartic acid*

When alkaloids of similar structures occur in the same plant species or genus, or even in the same plant family, it is likely that their biogenetic origins are similar or at least related. To form an idea of how closely alkaloids are related structurally, one can compare the nuclei and the substituents attached to them, and as we shall see, statistical methods have been successfully established for this purpose. First an insight is obtained into the interrelationships of a group of alkaloids from a study of their structures, from which some idea of the mode of biogenesis can be derived. One may then proceed to design and carry out feeding experiments on plants with suitably labeled precursors in order to test the bio-synthetic hypothesis experimentally.

6.1   BOTANICAL ORIGIN OF SOME IMPORTANT
1-BENZYLISOQUINOLINE ALKALOIDS

1-Benzylisoquinoline alkaloids and their derivatives occur in many plant families. The main groups shown in Scheme 1 have been isolated from species belonging to the following families:

Annonaceae (types I, III, VII)

Araceae (II)

Aristolochiaceae (II)

Berberidaceae (I, II, VII, VIII, IX)

Combretaceae (I)

Convolvulaceae (VII)

Euphorbiaceae (II, III)

Fumariaceae (II, VII, VIII, IX, X)

Hernandiaceae (I, II)

Lauraceae (I, II, III, VII)

Leguminosae (V)

Magnoliaceae (I, II)

Menispermaceae (I, II, III, IV, VII)

Monimiaceae (I, II, III)

Nymphaceae (I, II, III)

Papaveraceae (I, II, III, IV, VI, VII, VIII, IX, X)

Ranunculaceae (I, II, VII, VIII, IX)

Rhamnaceae (I, II)

Rutaceae (I, VII, VIII, X)

It is evident from this list that certain structural types occur frequently in the same plant family, and a closer study shows that some of these types are found in a single species, such as *Papaver somniferum* (I, IV, VII, VIII, IX) and *P. orientale* (II, III, IV, VII, VIII). The common occurrence of different structural types in the same plant provides an insight into the biogenetic relationships of the individual alkaloids.

6.2 STATISTICS FOR 1-BENZYLISOQUINOLINE ALKALOIDS

The group of isoquinoline alkaloids may be divided on the basis of structural similarity into various subgroups (see Section 5.1.7) which, however, may vary considerably in the number of members. Some statistics are presented in Scheme 1 concerning the position of the oxygen substituents on the nucleus of the more important subtypes, which in this sense are considered to be those with at least ten representatives. The examples given in the tables compiled by Raffauf [R 1], excluding bisisoquinoline bases, provide a uniform basis for the calculations. These conditions have been arbitrarily chosen for the statistical analysis. A different minimum number could have been selected, and moreover the restriction of the survey to a single literature source is questionable, but at least these conditions serve to provide unbiased statistical data. Other bases that could be added to this alkaloid group have been isolated subsequently, but it may be noted that if they were included in Scheme 1, no alteration in the overall picture would be produced.

Of the ten skeletal types shown in Scheme 1, only seven actually have an isoquinoline nucleus. The remainder, comprising the morphine, protopine, rhoeadine, and papaverrubine alkaloids, do not appear to belong to the same alkaloid class on the basis of their heterocyclic nuclei. Moreover, the number of carbon atoms in the skeleton may be either 16 or 17. Nevertheless a clear set of interrelationships can be established between all these skeletal types when the substitution patterns of the oxy functions indicated in Scheme 1 are taken into account.

In a certain sense the 1-benzyltetrahydroisoquinoline alkaloids appear to form a fundamental type as far as the whole group is concerned. They have the smallest number of rings with the smallest number of skeletal carbon atoms. All the examples in this group have an oxygen-containing function at C-4', and nearly all have similar substituents at C-6 and C-7. Apart from these three preferred positions, only C-3' with 42% substitution is worthy of note. If the pattern of substitution of the 1-benzyltetrahydroisoquinoline alkaloids is now compared with those of the other types given in Scheme 1, one finds good agreement for the most part:

Proaporphine alkaloids: oxy substituents mainly at C-1, C-2, C-10, and to a lesser extent at C-9

Erythrina alkaloids: C-15, C-16, C-3, and to a lesser extent at C-2

Berberine and protopine alkaloids: C-2, C-3, C-10, and to a lesser extent at C-9

Scheme 1 Commonly-occuring sub-groups of benzylisoquinoline alkaloids: a statistical survey of the pattern of oxygen substitution.[1]
Skeleton: $C_{16}N$

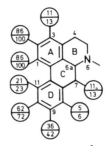

I. *1-Benzyl-1,2,3,4-tetrahydroiso-quinoline alkaloids* (26 representatives = 100%)[2]

II. *Aporphine alkaloids*[2] (86 representatives)

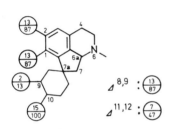

III. *Proaporphine alkaloids* (15 representatives)[3]

IV. *Morphine-type alkaloids* (15 representatives)[4]

IVA. *Morphine alkaloids without a ring ether group*[4] (10 representatives)

IVB. *Morphine alkaloids with a ring ether group* (5 representatives)

63

$\Delta^{1,6}$: $\frac{6}{50}$

$\Delta^{6,7}$: $\frac{10}{83}$

V. *Erythrina alkaloids* (12 representatives)

Skeleton: $C_{17}N$

VI. *Rhoeadine and papaverrubine alkaloids* (12 representatives)

VII. *Berberine alkaloids* (55 representatives)[5,6]

VIII. *Protopine alkaloids* (16 representatives)[6]

IX. *Phthalideisoquinoline alkaloids* (10 representatives)

X. *Benzophenanthridine alkaloids* (20 representatives)[2]

No. of examples with an
← *O*-containing group
← Overall percent

[1]The type of oxygen-containing substituent (OH, OCH_3, methylenedioxy, ketone) and the group attached to nitrogen [H, CH_3, $(CH_3)_2$, N-oxide] have not been specified throughout.
[2]Examples in which ring B is aromatic are also included.
[3]A carbonyl group occurs at position 10 in 12 proaporphine alkaloids.
[4]A carbonyl group occurs at position 7 in 9 morphine-type alkaloids.
[5]Alkaloids in which rings B and C are aromatic are also included.
[6]Alkaloids with a C-substituent at C-13 are included here and have not been treated separately.

In alkaloids of the phthalideisoquinoline, benzophenanthridine, and rhoeadine types, there are four carbon atoms that are substituted with more or less equal frequency, and which correspond to the positions of substitution in the 1-benzyltetrahydroisoquinolines. The aporphine and morphine alkaloids are the only ones that are strikingly different from the others. It is clear that this deviation cannot be accidental in view of the particularly large number of aporphine examples.

It will be observed that the number of carbon atoms substituted by oxygen is not only similar, but their location is also the same in every case. In particular, two of the four carbons that are most frequently substituted by oxygen are in vicinal positions to one another. In the majority of cases one carbon atom of this pair is substituted para (position 4) to an ethanamine residue, and thus the most important arrangement (with the exception of types IV-B, VI, and X) is as shown in the general formula 6/3.

6/3

On the basis of these general observations it should be possible to form a hypothesis concerning the molecular rearrangements that occur in the plant. This in turn should help to clarify the transformation of one skeletal type into another.

The proaporphine alkaloids (type III, Scheme 1) differ from the 1-benzyl-tetrahydroisoquinoline alkaloids (type I, Scheme 1) in having an extra bond between C-8 and C-1' in the latter type. As a result, type III has a spiro center and ring C is no longer aromatic. While the lower part of the molecule has an oxygen atom in the 4-position in relation to the spiro center, in the upper part there is an oxygen function in the ortho position to the new bond, which suggests that the formation of this bond takes place by a radical mechanism.

The aporphine alkaloids (type II, Scheme 1) also differ from the 1-benzyltetrahydroisoquinoline alkaloids (type I, Scheme 1) only in having an extra bond. C-8 is again involved, but this time it is linked to C-2′ of type I, and as a result both rings A and C remain aromatic. If we seek to explain the formation of this new bond through a radical mechanism as well, we run into difficulties. In the upper part of the molecule the new bond is again in an ortho position to an oxygen function, but in the lower part, neither the ortho (C-11) nor the para (C-9) position is very frequently substituted by oxygen. On the other hand the meta position has a relatively high rate of substitution (72%). We are thus led to the conclusion that the radical mechanism must be abandoned in favor of some other process, in view of the poor radical stabilization through a hydroxy group in a meta position. This other mechanism must be able to explain the distinctly lower frequency of oxygen substitution at C-10 in ring D, as compared to alkaloids of type I. A clue is given by the fact that substitution takes place at both positions 9 and 11, not just at one of them, as in alkaloids of types I or III.

A possible explanation would involve the reaction sequence:

$$I \longrightarrow III \longrightarrow II$$

The transformation of type I to type III may be regarded as a radical phenol oxidative coupling between C-8 and C-1′ (type I). The transformation of type III to type II could then result from an acid-catalyzed rearrangement involving rearomatiziation of ring D, as in Scheme 2 (dienone-phenol rearrangement).

Since alkaloids of type III are spiro compounds, two possibilities exist for this rearrangement (Scheme 2); the products differ in the location of the oxygen functions in the aporphine skeleton. Moreover, compounds of type III need not necessarily have a carbonyl group at C-10; the rearrangement could also take place if a C-10 hydroxyl group were present, in which case the latter group would be eliminated during the rearrangement (dienol-benzene rearrangement). These two types of reaction, involving two possibilities for the rearrangement and one for the elimination of the C-10 oxygen function, are in good agreement with the statistical results (see Scheme 1).

Similar considerations can be applied in the case of the biosynthesis of the morphine alkaloids (type IV, Scheme 1). Type IV-B is clearly formed from IV-A. All representatives of type IV-A have a C-7 oxygen group, which is presumably eliminated in the formation of type IV-B by attack of the C-4 hydroxyl on C-5 (Scheme 3), and a radical-type reaction similar to the transformation I → III provides a plausible mechanism for the cyclization involving the lower part of

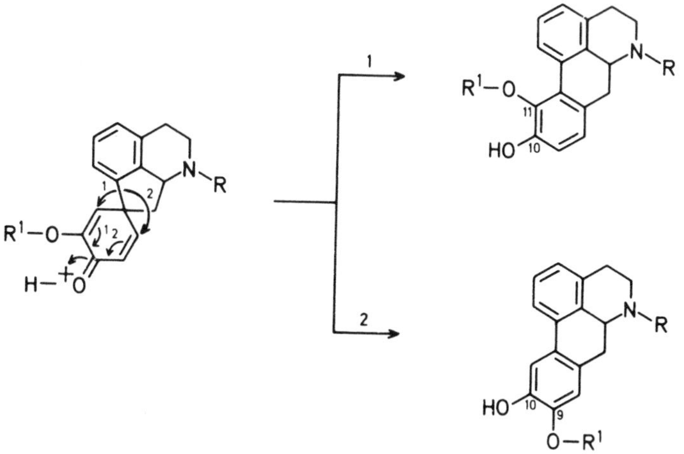

Scheme 2   Formation of aporphines from proaporphines.

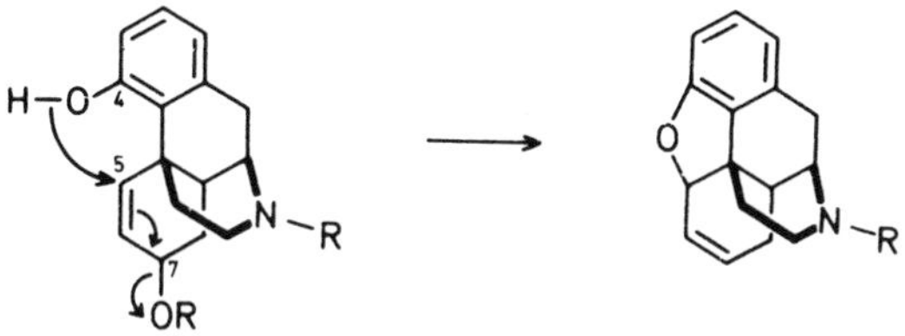

Type IVA   Type IVB

Scheme 3   Formation of ether ring in morphine-type alkaloids.

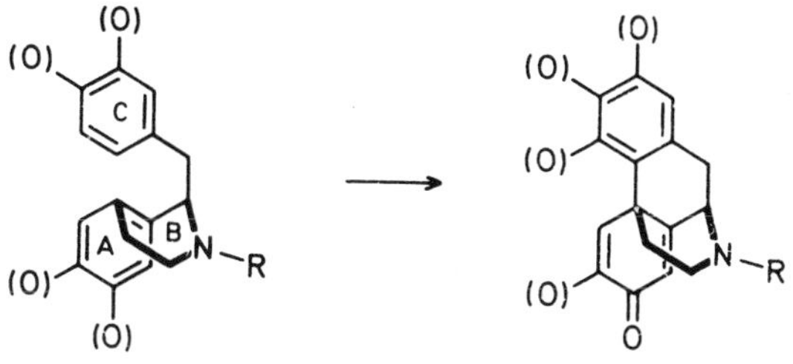

Type I   Type IVA

Scheme 4   Formation of morphine-type by cyclization.

the molecule (Scheme 4). However, the upper rings of alkaloids of type IV-A do not provide unequivocal evidence of a direct phenol oxidation from the statistical data in Scheme 1. The pattern of substitution in this ring resembles rather that of ring D of the aporphine alkaloids. Following the argument used in the latter case, one may deduce that an intermediate, 6-4, occurs as a stage in the biogenesis of the morphine alkaloids from alkaloids of type I. Type IV-B could then be formed by an acid-catalyzed rearrangement of 6/4. On the other hand, an analysis of the substitution pattern of individual alkaloids of type IV shows that *either* position 2 *or* 4 is substituted in addition to C-3. Thus a direct coupling (ortho or para) is possible in all cases without invoking the intermediate 6-4.

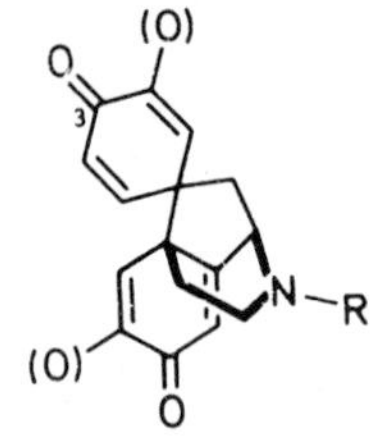

6/4

A striking feature of the *Erythrina* alkaloids (type V, Scheme 1) is the restriction of the oxygen substituents to a limited number of carbon atoms. From a comparison with representatives of the 1-benzyltetrahydroisoquinoline alkaloids, the following carbon atoms appear to correspond in the two types: 6 (type I) with 16 (type V); 7 with 15; 4' with 3; and 3' with 2. From this the sequence of reactions in Scheme 5 may be deduced.

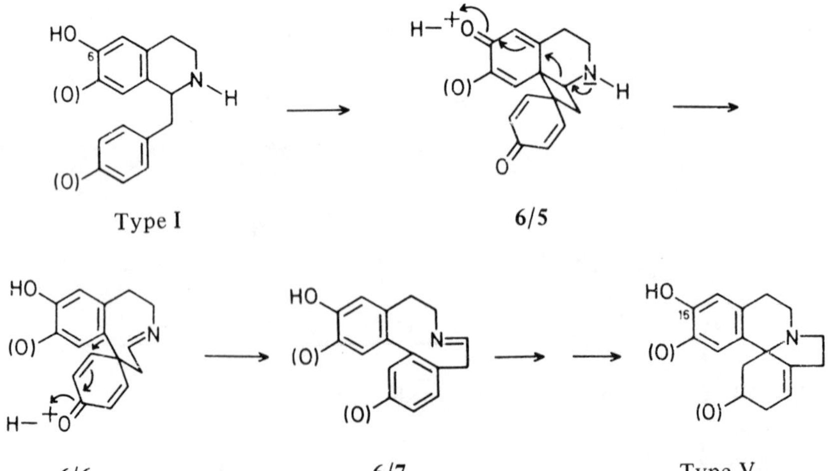

Type I   6/5

6/6   6/7   Type V

Scheme 5   Formation of *Erythrina* alkaloids from 1-benzyltetrahydroisoquinolines.

For the transformation of alkaloids of type I into the intermediate 6-5, a phenol oxidation with para-para coupling would be necessary. The two subsequent reactions in Scheme 5 will be recognized as a retro-Mannich reaction and a dienone-phenol rearrangement. The last stage of the rearrangement of 6-7 into type V alkaloids may be interpreted as a Michael-type cyclization involving oxidation of the lower aromatic ring. The sequence of reactions shown would lead to an alkaloid that corresponds to the observed pattern of substitution.

It is interesting that the isoquinoline alkaloids with a C_{17} skeleton also have a pattern of oxygen substitution corresponding to that of the C_{16} alkaloids. Thus alkaloids of type I may again be taken as providing the starting point. It will be observed that the berberine alkaloids (type VII, Scheme 1), the proto-pine alkaloids (type VIII, Scheme 1), the phthalideisoquinoline alkaloids (type IX, Scheme 1), and the rhoeadine alkaloids (type VI, Scheme 1) have an additional carbon atom attached at an ortho position to an oxygen function as compared to type I. This suggests that the same reaction is responsible for the attachment of the additional carbon atom in all cases. Type VII could be formed by a Mannich reaction involving the immonium form of a type I alkaloid, as shown in Scheme 6. Alkaloids of types VI, VII, VIII, and IX have an essentially

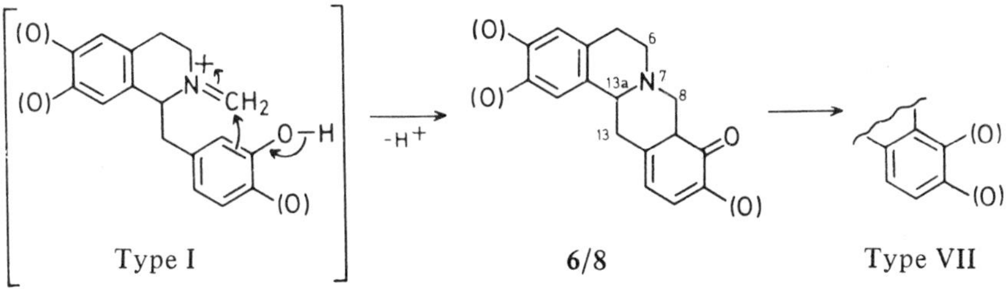

Type I   6/8   Type VII

Scheme 6   Conversion of 1-benzyltetrahydroisoquinolines to berberine-type alkaloids.[7]

higher frequency of oxygen substitution at C-3' (or C-5') as compared to type I. This may result from a process of natural selection whereby alkaloids of type I undergo the reactions shown in Scheme 6 only if they have an oxygen function (OH) at C-3' (or C-5'). The oxidative opening of the C-13a–N bond leads to the protopine type (VIII, Scheme 1), while the opening of the C-8–N bond followed by lactonization gives alkaloids of type IX (phthalideisoquinoline, Scheme 1). On the other hand if the C-6–N bond in type VII is opened and a bond is formed between C-6 and C-13, the result is a benzophenanthridine (type X) with a substitution pattern corresponding to the natural alkaloids[8]. Of the two

[7]The reaction pathway shown here corresponds to the results of feeding experiments on *Hydrastis canadensis*.
[8]This synthetic pathway was established by feeding experiments on *Chelidonium majus*.

tetrahydroisoquinoline nuclei in the berberine type, only the lower one remains as such in the benzophenanthridine bases.

From the substitution pattern in rings A and D of alkaloids belonging to the rhoeadine type a relationship to the phthalideisoquinoline alkaloids may be recognized, but it is not possible to deduce from this the biosynthetic pathway.

The various hypotheses that have been discussed concerning the formation of complex plant substances from simpler ones are based exclusively on a comparison of the ring nuclei and the frequency of oxygen substitution. Throughout this discussion one should not lose sight of the observation, which was made at the beginning of this section, that different plants can synthesize the same constituent in different ways. This consideration has been subordinated to a different objective, a study of the possible ways of formation of the main skeletal types of isoquinoline alkaloids.

The statistical method provides on the one hand clues as to possible biogenetic pathways, but on the other hand fine points concerning biosynthesis, which may be suggested by an individual alkaloid or by a group of bases from the same plant, tend to be averaged out and to disappear (e.g., considerations concerning the formation of type IVA). In spite of these various limitations and reservations, the procedure outlined above provides a good indication of possible biogenetic pathways, which must then be confirmed by experiment.

6.3   CHEMICAL REACTIONS INVOLVED IN THE TRANSFORMATION OF VARIOUS SUBGROUPS OF THE 1-BENZYLISOQUINOLINE ALKALOIDS [T 15]

The close relationship that clearly exists between individual subgroups of the isoquinoline series has stimulated efforts to find reactions that would allow the various types to be correlated with one another. Some of these reactions may be grouped as follows:

1 TRANSFORMATION OF 1-BENZYLTETRAHYDROISOQUINOLINES INTO APORPHINES Orientaline (6/9) is an alkaloid that belongs to the group of 1-benzyltetrahydroisoquinolines. In aqueous solution containing ammonium acetate and potassium ferricyanide, 6/9 is transformed in about 1% yield into the proaporphine alkaloid orientalinone (6/10, isolated from various species of *Papaver* including *P. orientale*). Sodium borohydride reduction of 6/10 gives the corresponding dienol, orientalinol (6/11) as a mixture of diastereomers. If this mixture is then allowed to stand at 20°C in aqueous methanolic hydrochloric acid overnight, a dienol-benzene transformation takes place that leads to the formation of the aporphine alkaloid isothebaine (6/12) [B 24] (see Scheme 7).

Although the overall yield from this reaction is quite modest, it nevertheless shows that such reactions can take place in a test tube under mild
reaction conditions such as occur in a plant.

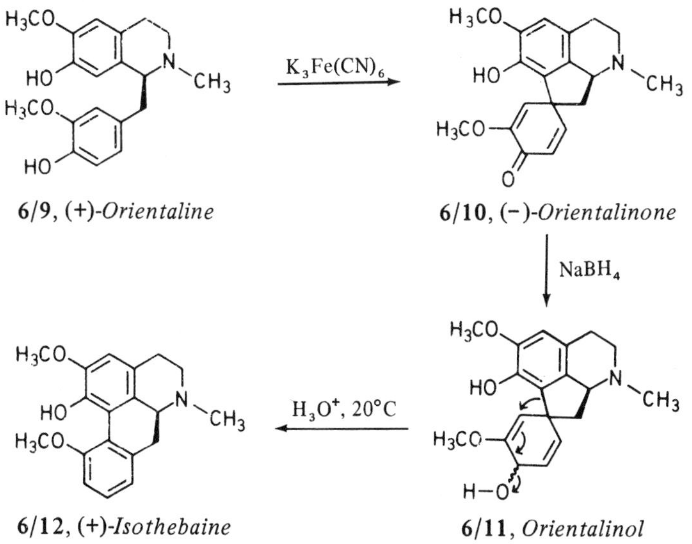

6/9, (+)-*Orientaline*　　　　　　　　6/10, (−)-*Orientalinone*

6/12, (+)-*Isothebaine*　　　　　　　6/11, *Orientalinol*

Scheme 7　Synthesis of (+)-isothebaine (6/12), from (+)-orientaline (6/9).

**2 TRANSFORMATION OF 1-BENZYLTETRAHYDROISOQUINOLINES INTO MOR-
PHINES** The 1-benzyltetrahydroisoquinoline 6/13 can be oxidized with
manganese dioxide to the dienone 6/14 (in a yield of only 0.024%). The
dienone can be reduced with sodium borohydride to 6/15 in the same way
as in the transformation of orientalinone (6/10) to isothebaine (6/12), then
6/15 in turn is converted in weakly acid medium at room temperature to
thebaine (6/16) (Scheme 8) [B 25] .

The detection of the very small quantity of thebaine produced was carried
out in an experiment using radioisotopes, in which labeled 6/13 was added,
and the solution containing 6/16 was diluted with inactive thebaine (dilution
analysis). The yield was determined from the radioactivity of the thebaine
that was isolated subsequently.

**3 TRANSFORMATION OF 1-BENZYLTETRAHYDROISOQUINOLINES INTO BER-
BERINE-TYPE ALKALOIDS** In the course of the formation of berberine-type
alkaloids from 1-benzyltetrahydroisoquinolines, an extra carbon atom must
be built in between the nitrogen and the lower benzene ring. Formaldehyde

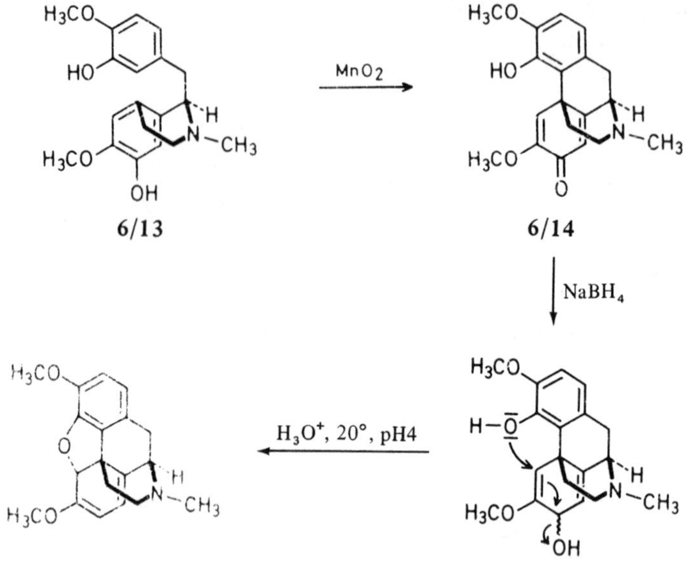

6/13   6/14   6/16, Thebaine   6/15

Scheme 8   Synthesis of thebaine **(6/16)** from the 1-benzyltetrahydroisoquino-line alkaloid reticuline **(6/13)**.

forms a possible source of this carbon atom, and its addition could take place through a Mannich-type reaction (Scheme 6). A necessary condition is that the lower ring contain a free phenolic hydroxyl group, which must be located ortho or para to the position where the new bond is to be formed. If the 1-benzyltetrahydroisoquinoline alkaloid (−)-norrecticuline **(6/17)** is allowed to stand at pH 6.3 in aqueous methanol containing sodium hydrogen carbonate and formalin at room temperature for a week, a mixture is obtained from which the two berberine-type alkaloids (−)-scoulerine **(6/18)** and (−)-coreximine **(6/19)** can be isolated in 43% and 24% yields, respectively (Scheme 9) [B 26]. The relative yields can be varied by adjustment of the pH.

This list of reactions could be further extended, but the examples will suffice to show that the transformation of simple naturally occurring isoquinoline derivatives into more complex substances also occurring naturally can take place in the laboratory under mild physiological-type conditions.

6.4   BIOGENESIS OF ALKALOIDS IN *PARAVER SOMNIFERUM*

We have already noted that the method of formation of a particular alkaloid is

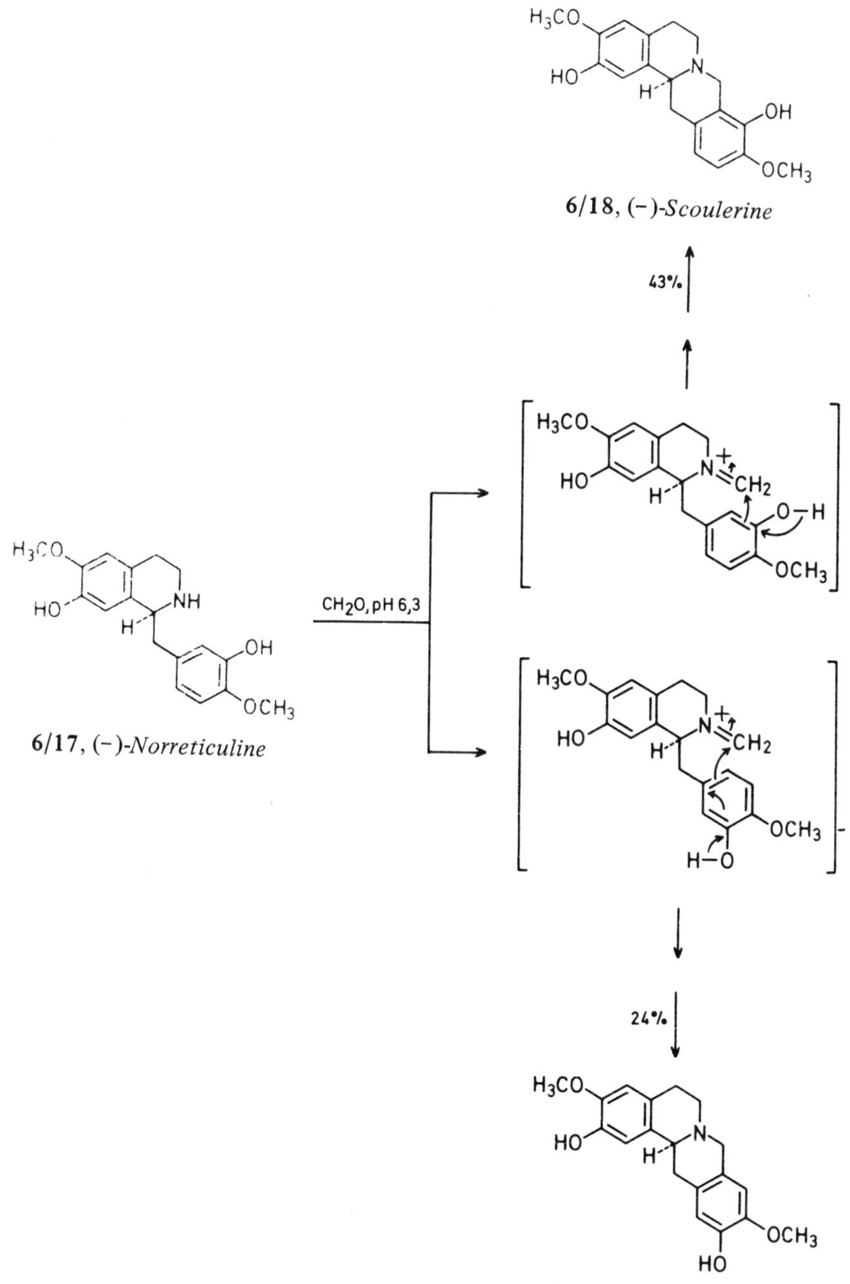

6/18, (−)-*Scoulerine*

43%

CH₂O, pH 6,3

6/17, (−)-*Norreticuline*

24%

6/19, (−)-*Coreximine*

Scheme 9   Synthesis of the berberine-type alkaloids scoulerine and coreximine from the 1-benzyltetrahydroisoquinoline norreticuline.

not necessarily the same in different plants. In this section we shall be considering the biogenesis of the alkaloids of one particular plant, *Papaver somniferum* (PAP). This example has been chosen because a very intensive study of this plant was made from a biogenetic point of view, and also because many subgroups of the isoquinoline alkaloids have been isolated from it. Studies on the biogenesis of the morphine alkaloids, particularly thebaine (6/16), codiene (6/28), and morphine (6/29) in *P. somniferum*, extended over many years and involved many individual experiments. In Scheme 10 the results of these investigations are summarized (see [M 34], [P 13], [S 41]).

Considerable interest was centered initially on the nature of the precursor of the isoquinoline alkaloids formed by *P. somniferum*. On feeding the amino acid tyrosine (6/20) as tyrosine-^{14}C-2 to the plant, the following labeled alkaloids were isolated: papaverine (6/23-^{14}C-1, ^{14}C-3), thebaine (6/16), codeine (6/28), and morphine (6/29-^{14}C-9, ^{14}C-16). It follows that two molecules of tyrosine are involved in the formation of both papaverine (6/23) and morphine (6/29). The formation of morphine from phenylalanine, which has no hydroxyl group, also takes place, but much less readily. If one feeds labeled norlaudanosoline (6/22) to the plant, it is aromatized and methylated to papaverine (6/23).

If dopa (6/21-^{14}C-2) is fed to the plant instead of tyrosine (6/20), the activity is located exclusively in carbon atom 16 of thebaine, codeine, and morphine. This shows that while tyrosine can form both halves of the 1-benzyltetrahydroisoquinoline alkaloids, dopa can only enter the amino half; moreover, the plant cannot transform dopa back into the corresponding phenylacetaldehyde.

By feeding the labeled intermediates indicated in Scheme 10 for the formation of morphine (6/29), the biosynthetic pathway of this alkaloid can be deduced as follows:

1 Codeine (6/28) is demethylated to morphine. The reverse process of methylation of morphine to codeine could not be detected.

2 The rate of incorporation of $^{14}CO_2$ and tyrosine (6/20-^{14}C-2) is at first greater with thebaine (6/16) than with codeine (6/28) and morphine (6/29). After a few days, morphine shows the greatest rate of formation, which means that in the scheme of biosynthesis thebaine is formed first, then codeine, and finally morphine.

3 Labeled thebaine is transformed into codeine and morphine, but neither codeine nor morphine is transformed into thebaine.

4 Labeled codeinone (6/26) is specifically transformed into codeine (6/28) and morphine (6/29).

5 Neopinone (6/27) is transformed in the plant into codeine (6/28).

6 Labeled salutaridine (6/14) is incorporated in good yield into the morphinane bases thebaine, codeine, and morphine. Moreover, the salutaridinols,

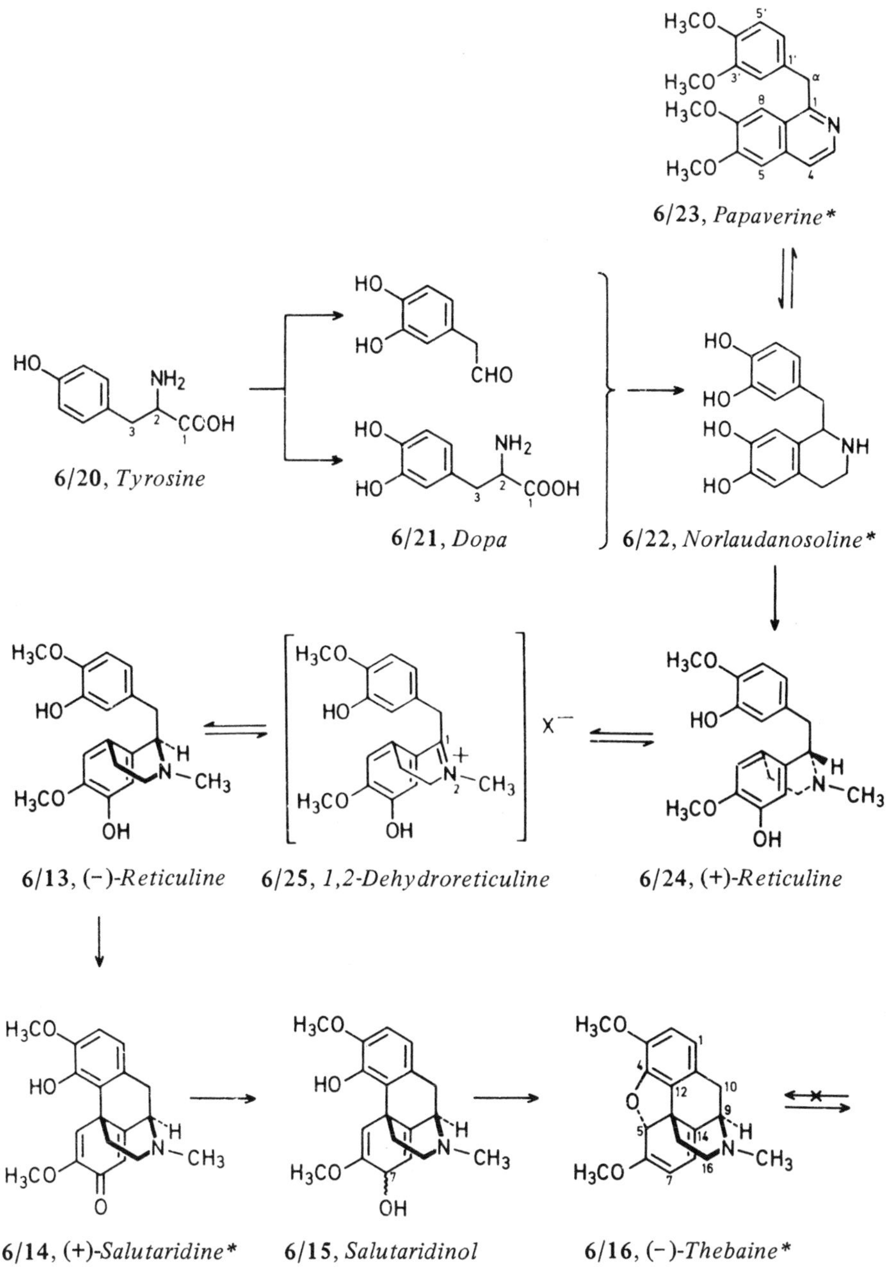

6/23, *Papaverine**

6/20, *Tyrosine*

6/21, *Dopa*

6/22, *Norlaudanosoline**

6/13, (−)-*Reticuline*   **6/25**, *1,2-Dehydroreticuline*   **6/24**, (+)-*Reticuline*

6/14, (+)-*Salutaridine**   **6/15**, *Salutaridinol*   **6/16**, (−)-*Thebaine**

*Isolated or detected in *P. somniferum*

Scheme 10   Biogenesis of morphine (**6/29**) in *Papaver somniferum*.

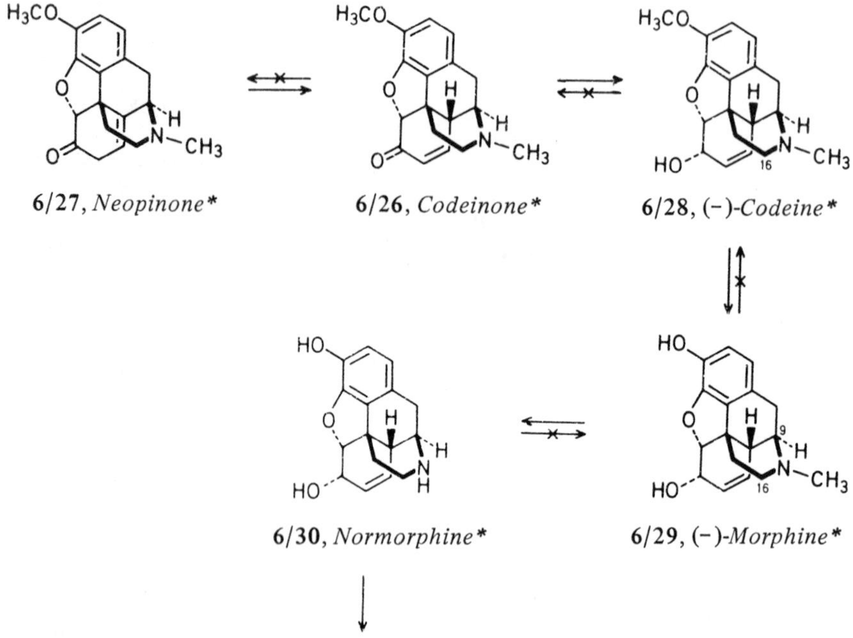

6/27, *Neopinone**   6/26, *Codeinone**   6/28, (-)-*Codeine**

6/30, *Normorphine**   6/29, (-)-*Morphine**

Non-alkaloid metabolites

Scheme 10 Continued.

which are epimeric at C-7, have also been shown to act as precursors for 6/16, 6/28, and 6/29.

7 The *O*- and *N*-methyl groups come mainly from methionine; formate is not incorporated so well.

8 The unmethylated 1-benzyltetrahydroisoquinoline derivative norlaudanosoline (6/22) is incorporated into papaverine (6/23), thebaine (6/16), codeine (6/28), and morphine (6/29) in higher yield than tyrosine. It can be concluded that 6/22 is closer than tyrosine to these alkaloids in the biogenetic pathway. On the other hand if *O,O,O,O*-tetramethylnorlaudanosoline is fed, the rate of incorporation into the morphinane alkaloids is very low. Evidently the methyl groups block further reaction. Neither 6/22 nor its tetramethyl ether corresponds in the *O*-methyl substitution pattern to thebaine (6/16), but reticuline (6/24 and 6/13) has the right number of methyl groups in the same positions as thebaine. Reticuline is incorporated into thebaine without demethylation.

9 *P. somniferum* degrades morphine (6/29) to normorphine (6/30), but the plant is not able to remethylate 6/30 to 6/29. Normorphine is apparently further degraded to nonalkaloidal metabolites.

10 A study has been made of a very interesting stereochemical aspect of this biosynthesis. The absolute configurations of the alkaloids (−)-thebaine, (−)-codeine, and (−)-morphine are given in Scheme 10, from which it appears that the true precursor of the morphinane alkaloids is (−)-reticuline (6/13), not (+)-reticuline (6/24). From this it might be deduced that only 6/13 lies on the biosynthetic pathway and not 6/24. Nevertheless both (+)- and (−)-reticuline are incorporated into the alkaloids. Feeding experiments with 6/24-1-T and 6/13-1-T establish the fact that tritiated (+)-reticuline gives a tritium-free morphine, while the (−)-reticuline, which corresponds to morphine in absolute configuration, is transformed into morphine with 60% of the original T-content intact. It may be concluded that between 6/24 and 6/13 in the biosynthetic scheme, an intermediate step occurs that is in equilibrium with epimers: if (+)-reticuline (6/24) is fed, it must be transformed via 6/25 (1,2-dehydroreticuline) into 6/13 and subsequent products, but (−)-reticuline (6/13) can give 6/14 directly by phenol oxidation. There is quite an appreciable loss of tritium on feeding 6/13, which is involved in the equilibration between 6/25 and 6/13. The incorporation has been verified by double labeling experiments involving an extra 3-T label.

The experimental results that have been described here are in agreement with the biosynthetic pathway set out in Scheme 10.

Two factors of decisive importance for the success of biogenetic experiments have not yet been mentioned:

1 Unequivocal results can be expected from plant feeding experiments only if the precursors are clearly labeled. The chemical synthesis of these precursor substances thus becomes very important. It is not possible due to limitations of space to go into details here.

2 The chemical degradation of the labeled metabolites is of no less importance. The total radioactivity is seldom determined. The information usually required is the exact location of the radioactive atom in the molecule, and the relative activity in comparison to the precursor. The degradation reactions must be unambiguous, otherwise the results may lead to false conclusions. Synthesis and degradation are usually the most time-consuming factors in biogenetic studies, and apart from this the feeding and incorporation experiments are subject to seasonal fluctuations.

In the above-mentioned scheme, carbon atoms 9 and 16 in morphine (6/29) are especially important, and a specific means for removing these two carbons is thus required. Two degradation reactions of morphine which need no comment and meet the demand for a specific degradation, are shown in Schemes 11 and 12.

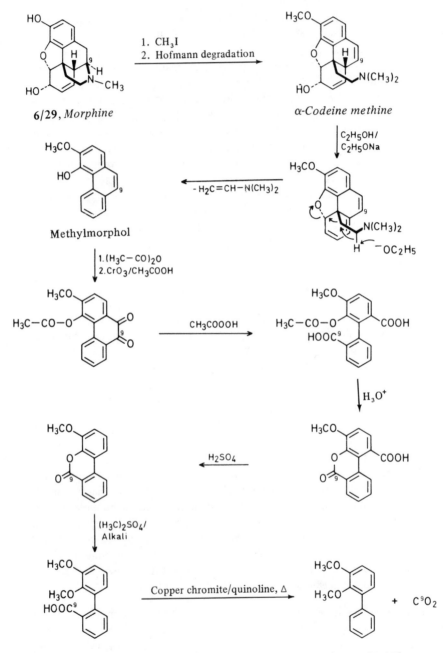

Scheme 11 Degradation of morphine (6/29) to obtain C atom 9 [B 27].

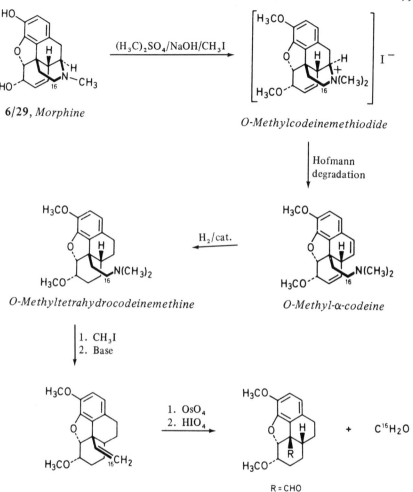

Scheme 12   Degradation of morphine (**6/29**) to obtain C atom 16 [B 28].

6.5   APORPHINE BIOGENESIS IN *PAPAVER ORIENTALE* [S 41]

Papaver orientale is especially suitable for the study of the biosynthetic pathway of aporphine alkaloids. The aporphine alkaloid isothebaine (**6/12**) and the proaporphines orientalinone (**6/10**) and dihydroorientalinone have been isolated from this plant, as well as the morphinane alkaloid thebaine (**6/16**). The latter base is of interest inasmuch as older theories considered that isothebaine was

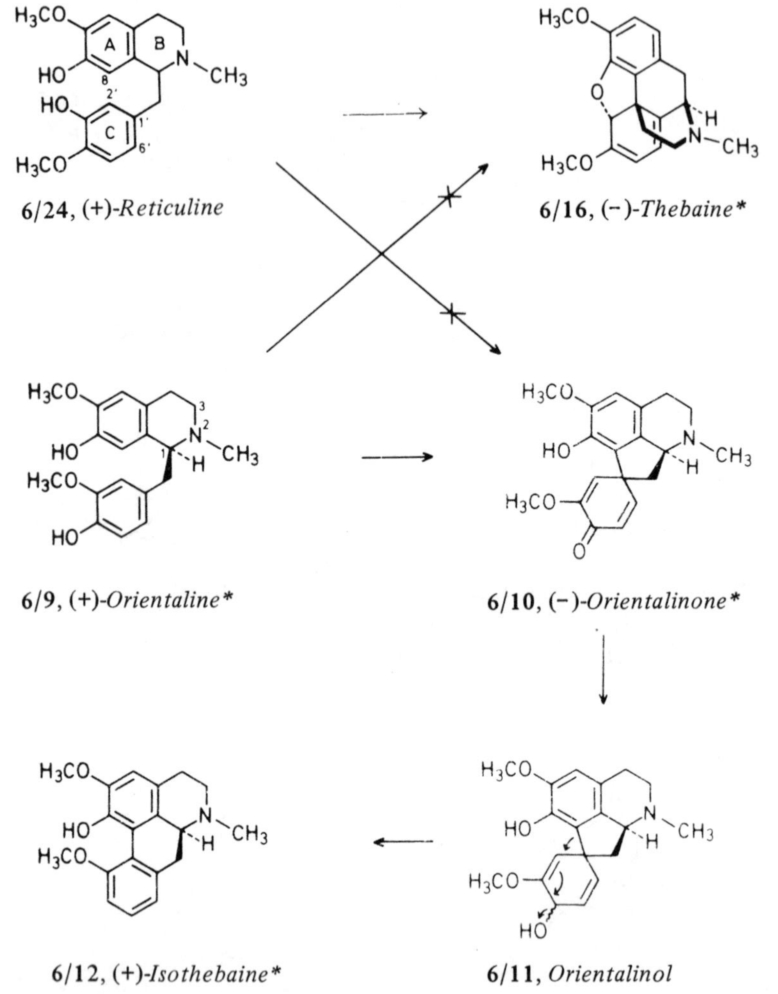

6/24, (+)-Reticuline

6/16, (−)-Thebaine*

6/9, (+)-Orientaline*

6/10, (−)-Orientalinone*

6/12, (+)-Isothebaine*

6/11, Orientalinol

*Isolated from P. orientale

Scheme 13   Aporphine biogenesis in Papaver orientale.

formed biosynthetically from thebaine, but this hypothesis is not substantiated by the labeling experiments described below.

   The actual results of the biosynthetic studies are summarized in Scheme 13.

1 Labeled orientaline (6/9), a 1-benzyltetrahydroisoquinoline alkaloid, produces active isothebaine (6/12) and inactive thebaine (6/16) when fed to

P. orientale. Specifically labeled methoxy groups in orientaline are retained as such in isothebaine, that is, there is no demethylation and remethylation on the pathway between orientaline and isothebaine.

2 Feeding experiments with tritiated (+)-orientaline have established that its incorporation takes place about 28 times better than with the corresponding (−)-enantiomer.

3 An interesting point is that reticuline (6/24) is incorporated only into thebaine (6/16) and not into isothebaine.

4 Orientalinone (6/10) has been shown to be specifically incorporated into isothebaine.

5 The various subsequent reactions of the 1-benzyltetrahydroisoquinoline alkaloids reticuline (6/24) and orientaline (6/9) are especially informative (see observations 3 and 4). The two alkaloids differ only in the substitution pattern in ring C, where methoxyl and hydroxyl groups are interchanged. They are thus isomers, and in orientaline the phenolic hydroxyl group of ring C is in a para position to the benzyl carbon atom, whereas in reticuline it is meta. A phenol oxidation takes place during the transformation orientaline → orientalinone. In this type of reaction, only a phenolic hydroxyl group in an ortho or para position can react; a methoxyl group or a meta hydroxyl group is ineffective. In ring A both reticuline and orientaline have a phenolic hydroxyl group that is ortho to the new C-C bond in orientalinone. From experimental evidence the plant cannot carry out demethylations at the orientaline (6/9) or reticuline (6/24) stage, and thus reticuline cannot be transformed into orientalinone. An alternative process leading to an aporphine skeleton which involves oxidative phenol coupling between C-2′ and C-8, or C-6′ and C-8, has so far not been observed in *P. orientale*, and it is not known whether this pathway is followed.

We have already discussed (Section 6.3) the further reactions of orientalinone (6/10) leading to isothebaine (6/12). Scheme 13, illustrating the biogenesis of the aporphine alkaloid isothebaine (6/12), has received support and confirmation from studies on other aporphines, such as the formation of stephanine (6/51) and aristolochic acid I (6/32) from orientaline (6/9), and of roemerine (6/33) from coclaurine (6/34) in various plants.

This review of two biosynthetic pathways will give an indication of the way in which such investigations are carried out, but it is clear that these two schemes cannot be regarded in any way as complete. A number of questions are left unanswered. For example, is isothebaine (6/12) degraded in the plant to a nitrogen-free product in the same way as morphine (6/29)? Is there a definite order followed in the methylation of the phenolic hydroxyl groups of norlaudanosoline (6/22), or is it not possible to determine this? Do other biosynthetic pathways exist in principle for aporphine alkaloids?

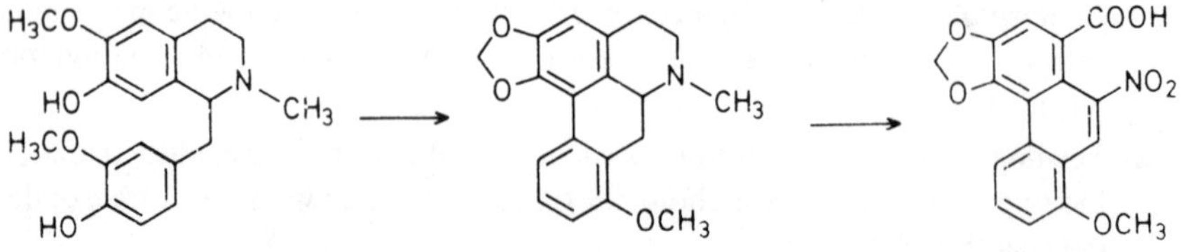

6/9 $\longrightarrow$ 6/31, *Stephanine* $\longrightarrow$ 6/32, *Aristolochic acid I*

6/34, *Coclaurine* $\longrightarrow$ 6/33, *Roemerine*

Aspects of Chemotaxonomy

The object of chemotaxonomy is the classification of plants or other living organisms on the basis of their chemical constituents. Thus chemotaxonomy provides further scientific data for the systematic botanist in addition to morphology, and the evidence derived from it should coincide with other data for the purpose of the arrangement and classification of plants. Chemotaxonomy could serve as a deciding factor in cases where other characteristics used in systematic botany are contradictory. In such cases it might be thought that chemotaxonomy would provide conclusive arguments for the clas,ification of plants. However, before we proceed to examine a specific example of chemotaxonomic reasoning, we must first of all clarify what we mean by chemical constituents of a plant and how we isolate them.

Plant constituents consist not only of alkaloids, but of a whole range of other substances such as flavonoids, coumarins, terpenes, sugars, and biogenic amines. They comprise substances of a very diverse nature, and we should go further and take into consideration the biogenesis of all these compounds in the particular plant we are dealing with. As we observed in the section on biogenesis, there are cases known in which the biosynthesis of the same constituent can take place in two different ways. Thus apart from the constituents themselves, their biogenesis in the plant can obviously provide chemotaxonomic evidence.

Because of the enormous range of plants that exist on earth, it is impossible for one individual or even for a whole institute to carry out studies on all their constituents, although this might be the best way of tackling the problem. One must rely on data provided in the chemical literature instead, and this is where the first difficulty comes in. In nearly every case an isolation is only reported when it gives a positive result, that is, publications usually describe the isolation of a particular plant constituent rather than reporting that a substance of a certain type could not be detected during the study of some plant. Such information, however, would considerably increase the value of chemotaxonomic studies.

The question as to whether a constituent is present or absent in a particular plant often depends on the quantity of the plant material available for the investigation. Clearly trace quantities of a compound cannot be detected in small amounts of the plant material. On the other hand if the plant under study is available in unlimited quantities—for example the apocynaceous plant *Catharanthus roseus* (*Vinca rosea*), from which the antileucemic bisindole alkaloids vinblastine and vincristine have been obtained—even trace constituents can be isolated in sufficient quantity to permit their structural elucidation. (It remains uncertain, of course, whether all the constituents of the plant have been obtained, or whether there might be other alkaloids present in still smaller quantities that have not yet been detected.)

Apart from these quantitative aspects, other factors are important in evaluating the constituents of a plant. The amount of a particular constituent present in different parts of the plant—bark, leaves, roots, flowers, and so on—can vary enormously, but often no information is given in the literature about which part of the plant has been investigated. Particulars concerning the season or the stage of development of the plant at the time of collection are even rarer. They are also of chemotaxonomic importance since the range of substances in a plant alters in the course of the vegetative period.

The description of the chemical properties of the basic constituents that have been isolated is often insufficient. It is often not possible to characterize the compounds completely due to lack of sufficient plant material; descriptions of this kind are nevertheless of very limited taxonomic value. Chromatographic and mass spectroscopic comparisons alone are not sufficient to establish the identity of two compounds. We already encountered several pairs of enantiomers among the alkaloids we have discussed, such as (+)- and (−)-quebrachamine, for which data on the optical properties, namely the $[\alpha]_D$ value or, better, the ORD curve, are essential for positive identification. On the other hand, a study of the optical properties of a compound is only meaningful when carried out on a completely pure substance. In the same plant compounds may occur whose $[\alpha]_D$ values are quite distinct, corresponding to the different types of chromophore of the indole alkaloids. Tiny quantities of impurities with a high specific rotation occurring in an alkaloid with a low specific rotation can lead to false interpretations. An example of the difficulties in evaluating data from the literature is provided by the alkaloid vincadifformine (Table 1). It has been shown by optical resolution that the alkaloid occurring in *Vinca difformis* is a racemate, but it is not clear whether both (±)- and (−)-vincadifformine occur in *Vinca minor*. In addition to the isomer with the highest specific rotation, there could also be partly racemized material in samples as usually prepared.

Table 1 $[\alpha]_D$ Values for Vincadifformine from Various Plants[a]

(+)-Vincadifformine

Plant Source	$[\alpha]_D$	Reference
Amsonia tabernaemontana	+600°	[Z 1]
Melodinus scandens	+526°	[M 35]
Rhazya stricta	+402°	[S 42]
Tabernaemontana riedelii	+185°	[C 13]
Vinca difformis, V. minor	± 0°	[D 3], [M 36]
Vinca minor	−540°	[P 14]
(−)-Tabersonine, synthesis	−600°	[Z 2]

[a]Solvent: C_2H_5OH, except for *V. difformis*: CH_3OH.

Having thus described the general basis of chemotaxonomy, we can now go on to examine a particular group of alkaloids from a taxonomic point of view. In doing so we must remember that the evaluation takes into account only one type of constituent, namely the alkaloids, and moreover, we can deal only with "positive" results, since usually these are the only ones that are recorded in the chemical literature [G 11].

7.1 CHEMOTAXONOMY OF THE PLUMERANE ALKALOIDS

We have seen that the total number of indole alkaloids of known structure amounts to about 1100, and these plant bases undoubtely constitute the largest single class of alkaloids. The most important subgroup of indole alkaloids is formed by the aspidospermine-type and related bases with about 250 representatives. In this subgroup the following types of nucleus are found (Scheme 14 gives an example of each):

3 Tetrahydrosecodine[1,2]

4 (−)-Quebrachamine

4Z (+)-(7R)-Dihydrocleavamine

[1]The number given for each skeletal type indicates the number of rings in the alkaloid; epoxides are not included.
[2]Alkaloids of types 3, 4, and 5 may form part of a biogenetic chain leading to other classes of indole alkaloids (see [K 5]).

5   (−)-Aspidospermine

5Z (+)-Pandoline

6A (−)-Pleiocarpine

6B Vindolinine

6C (−)-Cathovaline

6D (+)-Cimicidine

6E (−)-Hedrantherine

6Z (+)-Pandine

7A (+)-Kopsanol

7B Decarbomethoxyisokopsine

7C (−)-Fruticosine

7D (−)-Obscurinervine

The skeletal types 6C, 7B, and 7C are not dealt with in the discussion that follows since too few individual examples have so far been found. On the other hand the number of bisindole alkaloids with one or both halves consisting of these various types is relatively large with 38 members (see Section 10). The bisindole alkaloids will be referred to one or twice in the following discussion according to whether they have one or two bases of the above-mentioned types.

Scheme 14   Plumerane alkaloids: skeletal types.

Type	Structure[3]	Characteristic Example
3	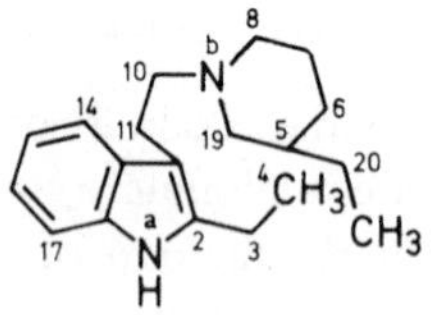	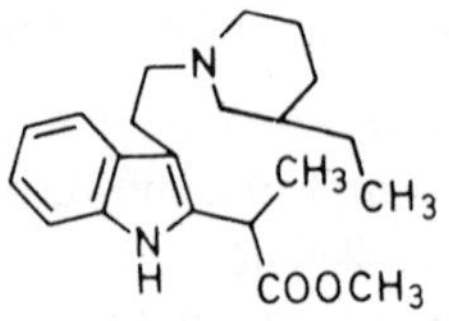 *Tetrahydrosecodine*
4	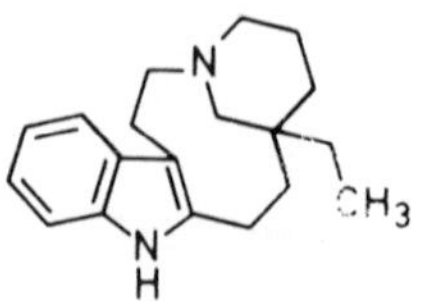	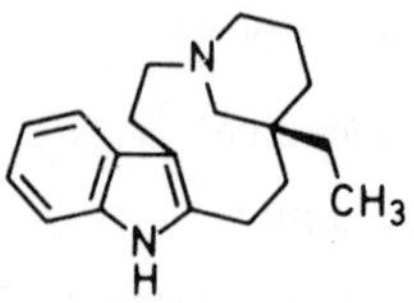 *(−)-Quebrachamine*

[3] A uniform system of numbering [H 14, H 15] has been adopted to facilitate comparison of the various structural types.

Scheme 14 Continued.

Type	Structure[3]	Characteristic Example
4Z	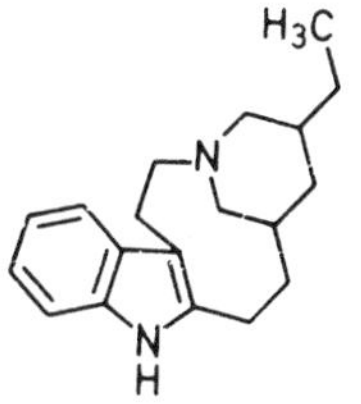	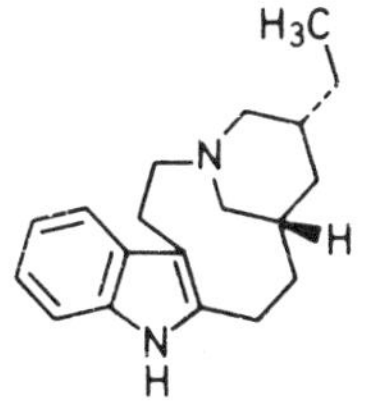 (+)-(7R)-Dihydrocleavamine
5	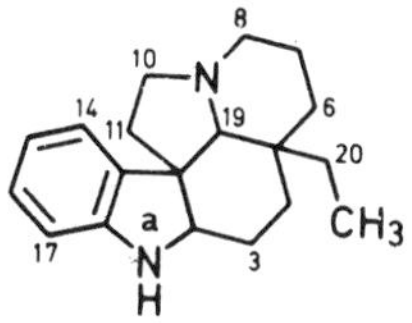	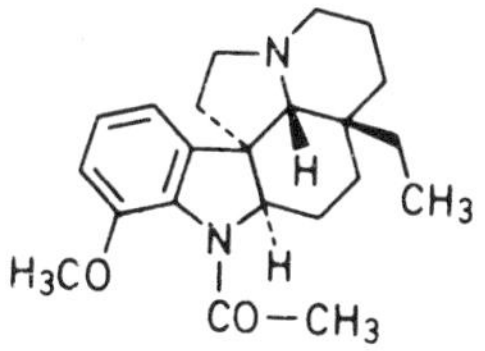
5Z	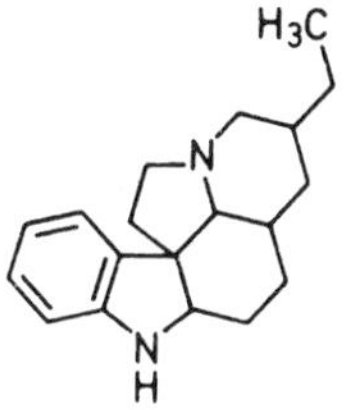	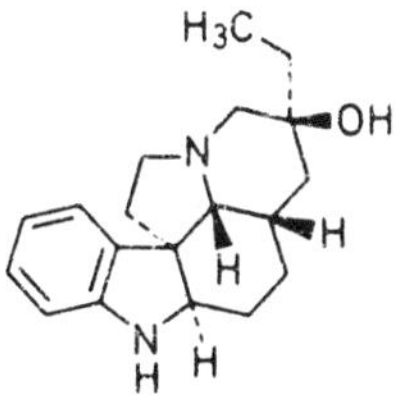 (+)-Pandoline
6A	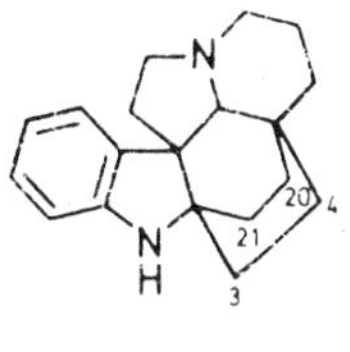	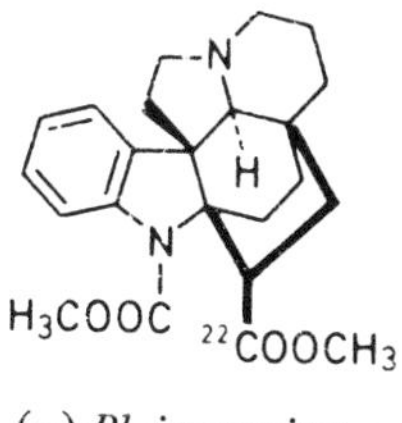 (−)-Pleiocarpine
6B	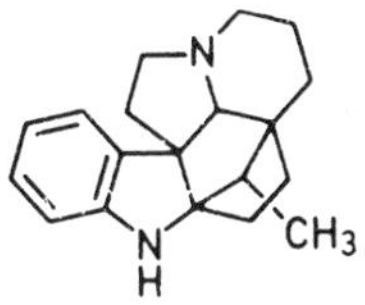	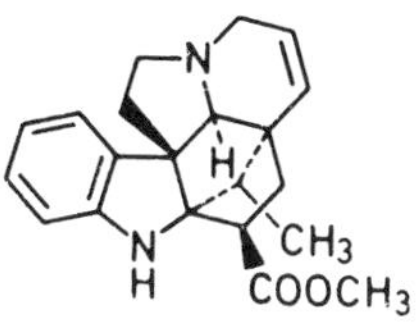 Vindolinine

Scheme 14 Continued.

Type	Structure[3]	Characteristic Example
6C	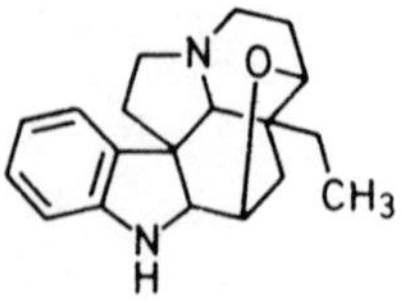	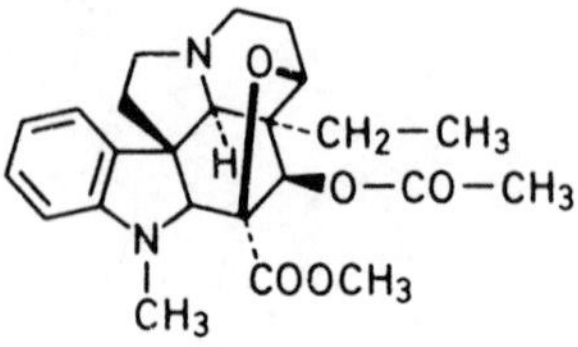 (−)-*Cathovaline*
6D	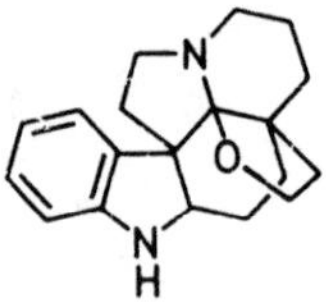	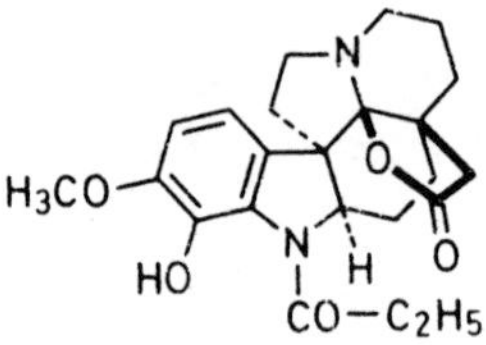 (+)-*Cimicidine*
6E	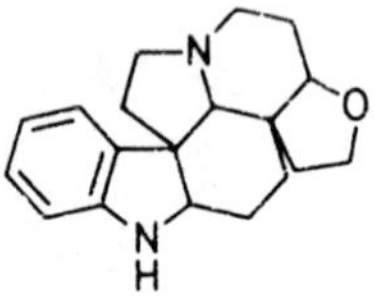	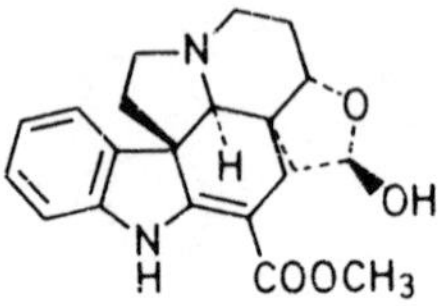 (−)-*Hedrantherine*
6Z	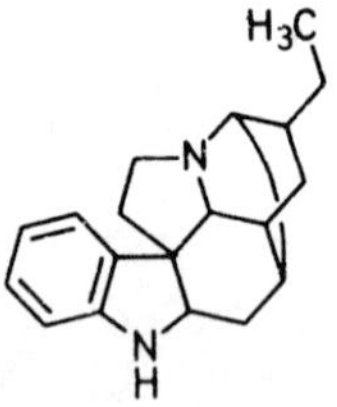	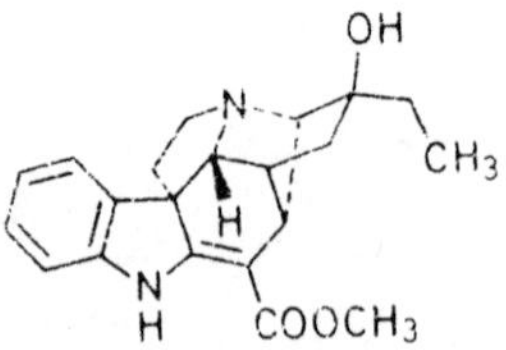 (+)-*Pandine*
7A	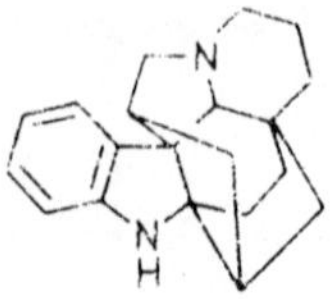	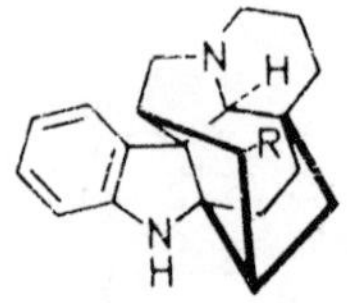 R = OH (+)-*Kopsanol*

88

Scheme 14   Continued.

Type	Structure[3]	Characteristic Example
7B	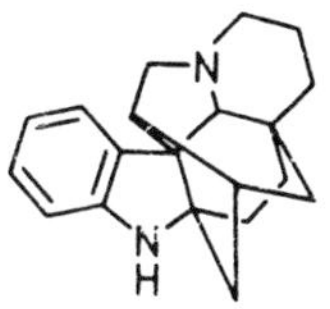	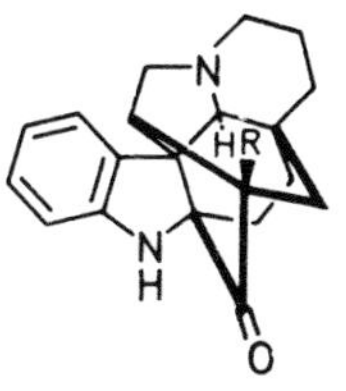 R = OH *Decarbomethoxyisokopsine*
7C	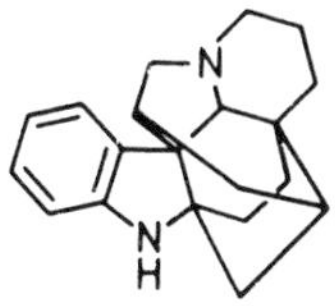	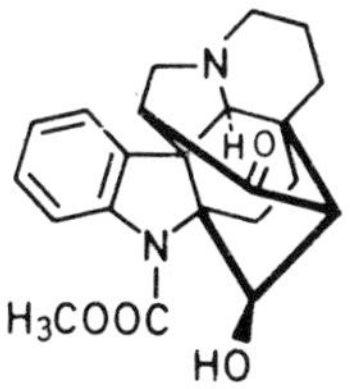 (−)-*Fruticosine*
7D	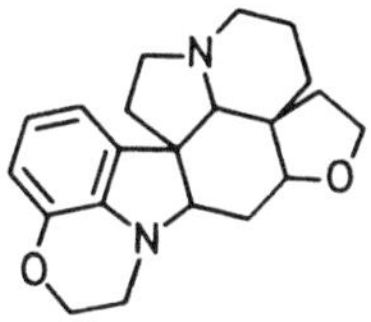	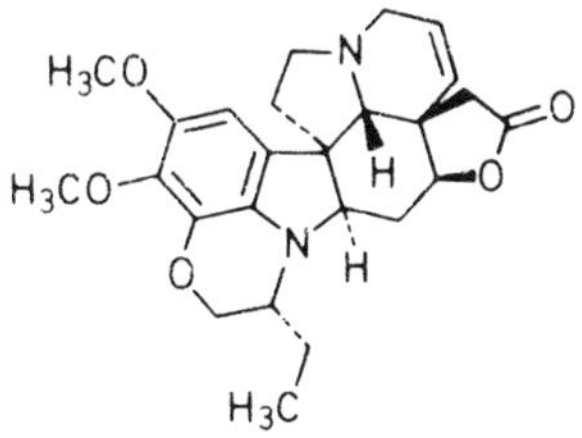 (−)-*Obscurinervine*

Bases of skeletal types 3-7 have been designated plumerane alkaloids in this book on account of their occurrence (see Table 2). and their close biogenetic relationship.

The intriguing question now arises as to whether a definite chemotaxonomic relationship exists between these alkaloids and the plants from which they have been isolated. The exclusive occurrence in the family Apocynaceae of the biogenetically related indole alkaloids, which have been grouped together in Scheme 14, provides a starting point for the investigation.

Other subgroups of indole alkaloids of a particular skeletal type are not necessarily so narrowly restricted in their occurrence. For instance, alkaloids of the yohimbane type have been isolated from plants belonging to the families Apocynaceae, Euphorbiaceae, and Rubiaceae; alkaloids of the heteroyohimbane type occur in the Apocynaceae, Loganiaceae, and Rubiaceae. There are, however, other subgroups of indole alkaloids which, like the plumerane alkaloids, occur in only one plant family. The advantage of the skeletal types 3-7 (Scheme 14), which have been selected, consists in the large number of individual members and the frequency with which they have been isolated (ca. 450 isolations[4] of 250 bases[5] from ca. 100 different plants). From the data available it should be possible to give a definite answer to the question that was posed at the outset as to whether chemical constituents can play a useful role in taxonomic classification.

7.2 CHEMOTAXONOMIC EVALUATION

In accordance with Table 2, the Apocynaceae can be subdivided into three subfamilies:

I   Plumieroideae

II   Cerberoideae

III Echitoideae

A large number of different alkaloids have been isolated from species distributed throughout all three subfamilies. Indole alkaloids, however, have been found in the subfamily Plumieroideae only. Apart from indole alkaloids (Section 5.1.2), the following other skeletal types have been isolated from the Apocynaceae: steroidal alkaloids (Section 5.6), monoterpenoid piperidine alkaloids of the skytanthine type (Section 5.1.3), and spermidine alkaloids (see Section 5.3), as well as the dimeric piperidine derivative carpaine[6] (Section 10.5). Skytanthine bases are the sole alkaloid type to have been found in the Cerberoideae. Steroidal alkaloids have been isolated only from plants of the subfamily Echitoideae and from those of the genus *Holarrhena*, which belongs to the first subfamily Plumieroideae. There is a striking relationship between this genus and the subfamily Echitoideae based on taxonomic criteria, which suggests that it should be reclassified.

---

[4]In this connection it is considered that one isolation consists in finding a certain alkaloid in a single plant species. Several isolations of the same alkaloid from the same plant by different authors is considered as only a single isolation. Bisindole alkaloids are counted singly or doubly according to the number of bases of the plumerane type they contain.

[5]For example, (+)- and (−)-quebrachamine are counted as two separate structures.

[6]The data on alkaloid occurrence are based on the following references: indole alkaloids in general: [H 14], [H 15], [H 16]; indole alkaloids of the plumerane type: [H 14], [H 15], [H 16], [G 11]; spermidine alkaloids: [H 20]; other types: [R 1].

Table 2 Botanical Classification of the Family Apocynaceae (K. Schumann, M. Pichon, and F. Markgraf)

I Plumieroideae

1. Carisseae

Parahancornia	Couma	Melodinus[a,b]	Carissa
Lacmellea	Hancornia[b]	Landolphia	Dictyophleba
Pacouria	Jasminochyla	Vahadenia	Clitandra
Carpodinus[b]	Willughbeia	Urnularia	Bousigonia
Leuconotis[b]	Cyclocotyla	Picralima[b]	Polyadoa[b]
Tetradoa	Hunteria[a,b]	Pleiocarpa[a,b]	

2. Chilocarpeae

Chilocarpus

3. Ambelanieae

Ambelania	Molongum	Rhigospira	Necocouma
Macoubea			

4. Tabernaemontaneae

Pagiantha[b]	Rejoua[a,b]	Ervatamia[a,b]	Pterotaberna
Hazunta[a,b]	Muntafara	Pandacastrum	Callichilia[a,b]
Ephippiocarpa	Hedranthera[a,b]	Tabernaemontana[a,b]	Peschiera[b]
Daturicarpa	Carvalhoa	Tabernanthe[a,b]	Schizozygia[b]
Stemmadenia[a,b]	Crioceras[a,b]	Calocrater	Voacanga[a,b]
Capuronetta[a,b]	Conopharyngia[a,b]	Pandaca[a,b]	Gabunia[b]

Table 2 Continued.

I Plumieroideae

5. **Alstonieae**

Craspidospermum[a,b]	Stephanostegia	Dyera	Kamettia
Gonioma[a,b]	Strempeliopsis	Plectaneia	Alstonia[a,b]
Tonduzia[b]	Winchia	Paladelpha	Bisquamaria
Blaberopus	Diplorrhynchus[b]	Aspidosperma[a,b]	Geissospermum[b]
Microplumeria	Rhazya[a,b]	Amsonia[a,b]	Lochnera[b]
Vinca[a,b]	Catharanthus[a,b]	Haplophyton[a,b]	Plumeria
Himatanthus	Holarrhena[c]		

6. **Rauvolfieae**

Cabucala[a,b]	Petchia	Rauvolfia[b]	Podochrosia
Alyxia	Lepinia	Lepiniopsis	Ochrosia[b]
Excavatia[b]	Vallesia[a,b]	Kopsia[a]	Rhipidia
Condylocarpon[b]	Anechites		

7. **Allamandeae**
Allamanda

II Cerberoideae

1. **Skytantheae**
Skytanthus[d]

2. **Thevetieae**

Cameraria	Cerberiopsis	Thevetia	Ahovai
Cerbera[d]			

III Echitoideae

1. Parsonsieae			
Chonemorpha[c]	Echites	Pachypodium	Parsonia
Prestonia	Rhabdadenia	Rhynchodia	Trachelospermum
Urechites			
2. Nerieae			
Adenium	Beaumontia	Christya	Funtumia[c]
Kibatalia[c]	Kicksia[c]	Paravallaris[c]	Malouetia[c]
Nerium	Pottsia	Roupellina	Strophanthus
Vallaris	Wrightia[c]		
3. Ecdysanthereae			
Anodendron	Apocynum	Chavannesia	Cleghornia
Odontadenia	Urceola		
4. Ichonocarpeae			
Elytropus	Forsteronia	Ichnocarpus	Mandevilla
Dipladenia	Macrosiphonia	Parameria	Oncinotis[e]

The following alkaloids were found to date:

[a]Plumerane alkaloids.

[b]Indole alkaloids and related bases, except plumerane type.

[c]Steroidal alkaloids (see Section 5.6).

[d]Piperidine alkaloids (see Section 5.1.3).

[e]Spermidine alkaloids (see Section 5.3).

As mentioned previously alkaloids of the plumerane type, that is of skeletal types 3-7 (Scheme 14), have been isolated or detected exclusively in members of the family Apocynaceae. It is remarkable that these bases occur only in the Plumieroideae (from which the name of this alkaloid group is derived). The Plumieroideae are further divided into seven tribes. So far, indole alkaloids of the plumerane type have been detected in only four of them.

The fact that these types of indole alkaloids occur exclusively in the subfamily Plumieroideae of the Apocynaceae provided the motive for this chemotaxonomic investigation. The question arises whether additional chemotaxonomic evidence can be derived from the chemical and structural criteria, and whether the botanical classification is in agreement with it.

7.2.1 Skeletal Types

Table 3 gives the distribution of the plumerane alkaloids in four tribes of the subfamily Plumieroideae. Skeletal types 4Z, 5Z, and 6Z, comprising eight bases, all of which occur in the genera *Pandaca* and *Capuronetta* (tribe 4, Tabernaemontaneae), are not included; for their separate treatment see the end of this section.

The main alkaloids under consideration are represented by skeletal types 4 and 5, both of which occur in the four tribes listed in the table. This is understandable since both skeletal types play a central role in a biogenetic sense. The types with six rings, which are derived from type 5 and have more complicated structures compared with type 5, do not occur in all four tribes. Type 6A, which occurs in three tribes, forms the starting point for the formation of the skeletal type 7A, at least as far as *in vitro* experiments show. It is noteworthy in this connection that alkaloids of type 7A are only found in tribes in which type 6A also occurs. Type 7A is moreover the starting point for type 7B and 7C skeletons, as is also shown by *in vitro* experiments. Alkaloids with the 7B and 7C nuclei, clearly the most complicated alkaloid types in this series, occur exclusively in the tribe Rauvolfieae, which also contains types 7A and 6A. Furthermore it can be seen from Table 3 that there are tribes which form only one higher skeletal type, which is characteristic of them. The relatively frequent skeletal type 6E, which has been isolated 74 times, is characteristic for plants of the tribe Tabernaemontaneae. So far it has not been detected in other tribes. Type 7D is far less common, but it is equally restricted, in this case to the Alstonieae, although it cannot be taken as characteristic of the tribe since it has been isolated from only two species so far. A detailed study of the individual genera, arranged in tribes in Tables 4-7, confirms the results set out in the summary presented in Table 3.

The three genera studied so far in the tribe Carisseae, of which eight species have been investigated, show an essentially uniform picture (Table 4). Starting

Table 3 Distribution of Plumerane Alkaloids in the Tribes of the Subfamily Plumieroideae[a]

Tribes	Species Investigated	Skeletal Types (see Scheme 14)									7B	
		3	4	5	6A	6B	6C	6D	6E	7A	7C	7D
1. Carisseae	8	—	3(3)	13(15)	19(45)	8(8)	—	—	—	3(4)	—	—
4. Tabernaemontaneae	37	1(1)	5(13)	19(28)	—	—	—	—	19(74)	—	—	—
5. Alstonieae	58	11(27)	14(27)	87(165)	20(27)	3(6)	3(4)	14(18)	—	4(13)	—	5(5)
Aspidosperma spp. included therein	[36]	—	3(9)	43(80)	11(14)	—	—	9(13)	—	4(13)	—	5(5)
6. Rauvolfieae	8	—	1(1)	9(11)	3(3)	—	—	5(5)	—	2(4)	3(3)	—
Total	111											
Subtotal isolations	510	28	44	219	75	14	4	23	74	21	3	5
Total isolations	510	28	44	219			190				29	
Percent isolations	100	5.5	8.6	42.9			37.3				5.7	

[a]The figures give the number of structurally different alkaloids; those in parentheses indicate the number of isolations.

Table 4 Distribution of Alkaloids according to Skeletal Type in the Tribe Carisseae[a]

Genus	Species Investigated	Skeletal Types (see Scheme 14)									7B	
		3	4	5	6A	6B	6C	6D	6E	7A	7C	7D
Melodinus	4	—	2(2)	8(10)	6(11)	5(5)	—	—	—	—	—	—
Hunteria	1	—	—	3(3)	6(6)	—	—	—	—	—	—	—
Pleiocarpa	3	—	1(1)	2(2)	13(28)	3(3)	—	—	—	3(4)	—	—

[a]The figures give the number of structurally different alkaloids; those in parentheses indicate the number of isolations.

Table 5 Distribution of alkaloids in the Tribe Tabernaemontaneae[a]

| Genus | Species Investigated | Skeletal Types (see Scheme 14) | | | | | | | | | 7B | | |
		3	4	5	6A	6B	6C	6D	6E	7A	7C	7D
Rejoua	1	—	—	—	—	—	—	—	1(2)	—	—	—
Ervatamia	1	—	—	1(1)	—	—	—	—	—	—	—	—
Hazunta	1	—	—	2(2)	—	—	—	—	2(6)	—	—	—
Callichilia	2	—	—	—	—	—	—	—	10(16)	—	—	—
Hedranthera	1	1(1)	1(1)	—	—	—	—	—	—	—	—	—
Tabernaemontana	7	—	1(3)	10(10)	—	—	—	—	10(16)	—	—	—
Conopharyngia	2	—	1(1)	1(1)	—	—	—	—	1(2)	—	—	—
Capuronetta	1	—	—	—	—	—	—	—	—	—	—	4Z,5Z,6Z
Pandaca	5	—	1(1)	1(1)	—	—	—	—	—	—	—	4Z,5Z,6Z
Tabernanthe	1	—	1(1)	—	—	—	—	—	—	—	—	—
Stemmadenia	3	—	1(1)	1(3)	—	—	—	—	—	—	—	—
Crioceras	1	—	2(2)	2(2)	—	—	—	—	1(2)	—	—	—
Voacanga	11	—	3(3)	6(7)	—	—	—	—	12(46)	—	—	—

[a]The figures give the number of structurally different alkaloids; those in parentheses indicate the number of isolations.

Table 6 Distribution of Alkaloids in the Tribe Alstonieae[a]

Genus	Species Investigated	\ Skeletal Types (see Scheme 14)											
		3	4	5	6A	6B	6C	6D	6E	7A	7B	7C	7D
Craspidospermum	2	—	—	1(2)	1(2)	—	—	—	—	—	—	—	—
Gonioma	1	—	2(2)	2(2)	1(1)	—	—	—	—	—	—	—	—
Alstonia	1	—	—	5(5)	3(3)	—	—	—	—	—	—	—	—
Rhazya	2	10(19)	2(2)	4(4)	—	—	—	—	—	—	—	—	—
Amsonia	3	3(8)	5(6)	6(8)	—	—	—	—	—	—	—	—	—
Vinca	7	—	7(7)	19(24)	6(7)	2(2)	—	—	—	—	—	—	—
Catharanthus	4	—	—	21(38)	2(2)	1(4)	3(4)	—	—	—	—	—	—
Haplophyton	1	—	—	—	—	—	—	5(5)	—	—	—	—	—
Aspidosperma													
1. *Macrocarpa*	3	—	—	—	—	—	—	—	—	5(13)	—	—	—
2. *Nitida*	4	—	—	4(8)	—	—	—	1(1)	—	—	—	—	—
3. *Nobiles*	10	—	1(3)	22(29)	—	—	—	8(13)	—	—	—	—	—
4. *Polyneura*	6	—	1(2)	21(24)	—	—	—	—	—	—	—	—	5(5)
5. *Pyricolla*	8	—	—	6(7)	11(14)	—	—	—	—	—	—	—	—
6. *Quebrachines*	2	—	3(4)	10(10)	—	—	—	—	—	—	—	—	—

[a]The figures give the number of structurally different alkaloids; those in parentheses indicate the number of isolations.

Table 7 Distribution of Alkaloids in the Tribe Rauvolfieae[a]

Genus	Species Investigated	Skeletal Types (see Scheme 14)										
		3	4	5	6A	6B	6C	6D	6E	7A	7B 7C	7D
Cabucala	2	—	—	2(3)	—	—	—	—	—	—	—	—
Vallesia	2	—	1(1)	6(8)	—	—	—	5(5)	—	—	—	—
Kopsia	4	—	—	—	3(3)	—	—	—	—	2(4)	3(3)	—

[a]The figures give the number of structurally different alkaloids; those in parentheses indicate the number of isolations.

from skeletal types 4 and 5, the main types formed are 6A and 6B; only *Pleiocarpa* is able to transform 6A into 7A (see Scheme 15).

The tribe Tabernaemontaneae also presents a straightforward picture. It forms only the common skeletal types 4 and 5 and the special nucleus 6E, which, as we have seen, occurs exclusively in plants of the Tabernaemontaneae. The occurrence of type 6E is restricted to one genus, but it is widely scattered throughout that genus. Types 4Z, 5Z, and 6Z, which have been isolated only recently and exclusively from *Pandaca* and *Capuronetta* species, occupy a special position. They differ from types 4 and 5 in having the ethyl group at position 7 instead of the usual position 5. The tetracyclic base cleavamine, which has the 4Z nucleus and forms the basis for the two other Z types, has long been known as a degradation product of the *Iboga* alkaloid catharanthine. In this connection it should be noted that the biogenesis of the Z types is associated with the *Iboga* alkaloids and not with the plumerane-type bases. This inference is confirmed by the fact that *Iboga* alkaloids have been isolated from eleven genera of the Tabernaemontaneae. Preliminary considerations suggest that the *Iboga*-type nucleus is quite suitable as an additional characteristic for genera of the tribe Tabernaemontaneae.

(+)-*Catharanthine*   *Cleavamine*

The most intensively studied tribe is Alstonieae with a total of 58 species from nine genera which have been chemically characterized. As far as the distribution of the skeletal types in the individual genera is concerned, this tribe is less uniform. *Haplophyton* and *Aspidosperma* in particular synthesize type 6D but not 6A, which is significant as far as other Alstonieae genera are concerned; *Amsonia* and *Rhazya* likewise do not form 6A, but produce type 3 instead, which is probably formed from 4 and 5 but which could also arise equally well from other indole alkaloid types (e.g., stemmadenine, *Iboga*).

A chemotaxonomic evaluation of the fourth tribe Rauvolfieae is not possible at present, since 23 different alkaloids have been isolated from the eight species belonging to three genera which have so far been investigated. These 23 compounds are distributed among seven skeletal types, and the statistical material available is too little in this case. There is no botanical relationship between *Kopsia*, *Pleiocarpa*, and the subgenus *Macrocarpa* of the *Aspidosperma*, which one might expect from the common occurrence of skeletal type 7A.

7.2.2 Absolute Configuration

In a search for more precise criteria for chemotaxonomy, a study has been made of the absolute configurations of individual alkaloids. Unfortunately only about 50% of the alkaloids isolated can be used for this purpose, because in many cases the necessary data are not available from the literature, or the absolute configuration is not yet known.

The alkaloids so far considered can have several chiral centers in the skeleton. Neglecting type 3 (Scheme 14), the number of these centers can vary between one (type 4) and six (type 7A). (Chiral centers other than those in the skeletal types are not considered.) As one might expect from the large number of fused ring systems, individual centers show a mutual dependence in orientation. A particular difficulty in the nomenclature of different skeletal types that have the same absolute configuration consists in the fact that centers which have the same number in the enumeration system are not chiral for all the members of that type. In types 5, 6, and 7, centers 12 and 19 are always chiral, but in some cases center 5 is not (e.g., for type 6A). On the other hand in type 4, centers 12 and 19 are not chiral, but center 5 is exclusively so. For a uniform and rapid indication of the absolute configuration of alkaloids of types 3-7 we shall use center 5, but in the exceptional cases referred to above, center 12 will be used instead; it always has the reverse configuration as a result of cyclization. Thus type 5 will be designed 5-α if center 5 has an α-configuration. Contrary to common usage, α and β refer here to the absolute configuration. Unfortunately R,S-nomenclature cannot be used for our purposes, since different substituents can lead to a change of priority for the same absolute configuration, which will result in the opposite R,S designation. For chemotaxonomic evaluation, a uniform method of designation for the same absolute configuration is desirable.

5 α   5 β

In Schemes 15-18 the skeletal types are grouped together according to their absolute configurations for each individual tribe. The arrows between the formulas indicate the most likely biogenetic path from the fundamental type, as discussed previously (see [K 5] for a more detailed discussion). Relationships not covered in [K 5] (skeletal types 6D, 6E, and 7D) can readily be explained as a result of simple chemical reactions.

The absolute configuration of alkaloids of type 3 cannot at present be established from the published data available in the literature. Types 4 and 5,

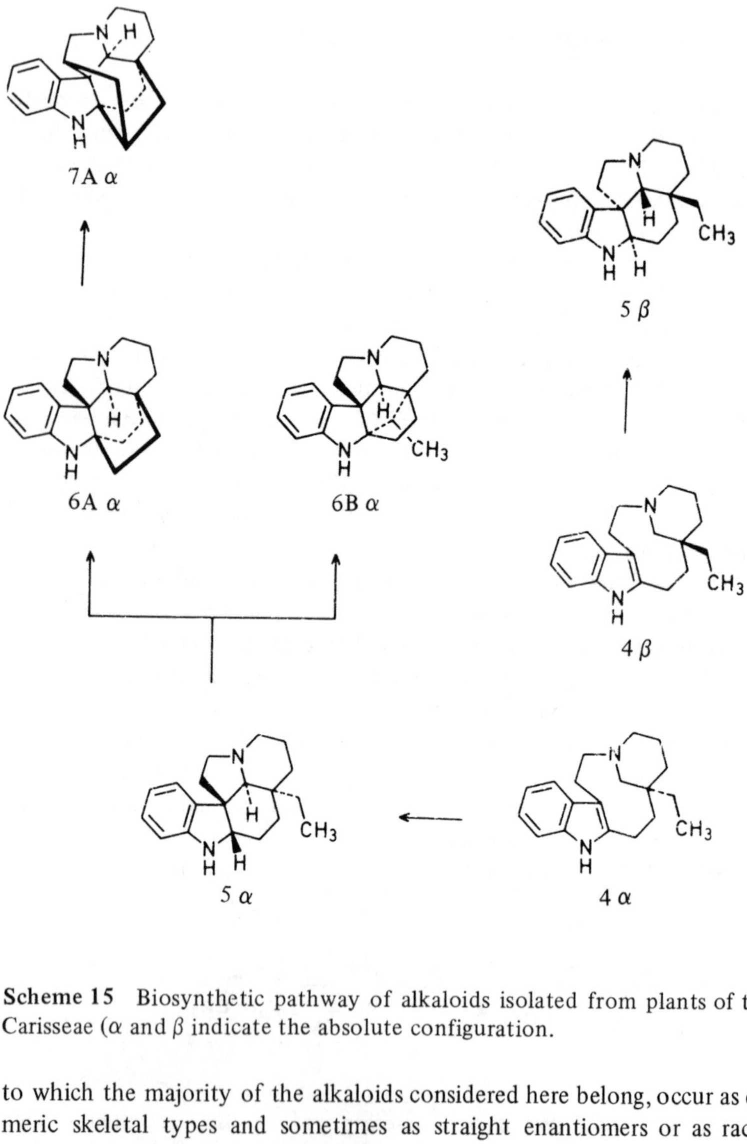

7A α

5 β

6A α   6B α

4 β

5 α   ←   4 α

Scheme 15 Biosynthetic pathway of alkaloids isolated from plants of the tribe Carisseae (α and β indicate the absolute configuration.

to which the majority of the alkaloids considered here belong, occur as enantiomeric skeletal types and sometimes as straight enantiomers or as racemates, such as vincadifformine (see Table 1) and quebrachamine (see Scheme 15). Types 6A, 6B, and 6E and 7A so far have been found with only one absolute configuration, and we shall designate them α.[7] On the other hand types 6D and 7D have the β-configuration exclusively. From this observation it follows

[7]Skeletal types 7B and 7C which occur in the tribe Alstonieae have only the α-configuration (see Scheme 14).

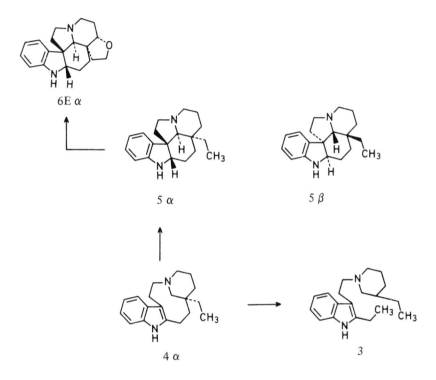

Scheme 16 Biosynthetic pathway of alkaloids isolated from plants of the tribe Tabernaemontaneae (α and β indicate the absolute configuration).

that with increasing complexity, the formation of a skeleton becomes more stereospecific. The meaning of complexity in this connection is clear: alkaloids with six or more rings occur with one absolute configuration only.

Plants of the tribe Carisseae (Scheme 15) produce types 4 and 5 in both configurations, but the more complex types have only the α-configuration. The tribe Tabernaemontaneae (Scheme 16) behaves similarly; the fundamental type 4-β so far has not been found. The tribe Alstonieae (Scheme 17), which has been most intensively studied, shows two independent biosynthetic pathways. One starts from 5-β and leads to types 6D and 7D, and the other includes the formation of types 6A, 6B, and 7A from 5-α. It may be inferred for plants of this tribe that the formation of skeletal types takes place stereospecifically as in the previous cases. In the tribe Rauvolfieae (Scheme 18), where the isolation of type 4-α would be expected, both biosynthetic pathways occur as in the previous case (β: 6D; and α: 6A and 7A). The genus of the Apocynaceae that has the most species and has been most intensively studied is *Aspidosperma*.

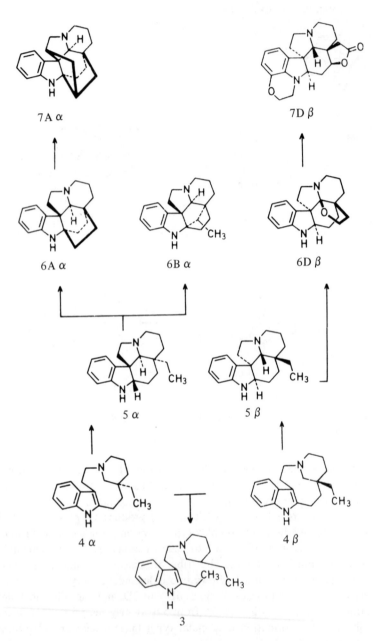

Scheme 17 Biosynthetic pathway of alkaloids isolated from plants of the tribe Alstonieae (α and β indicate the absolute configuration).

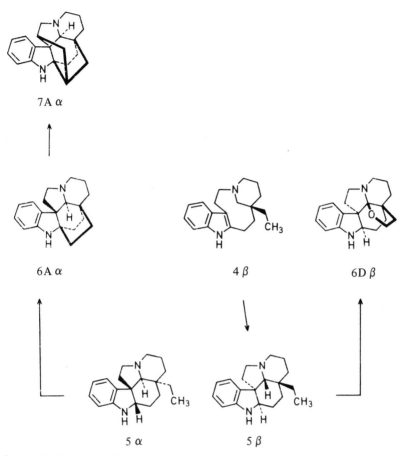

Scheme 18 Biosynthetic pathway of alkaloids isolated from plants of the tribe Rauvolfieae (α and β indicate the absolute configuration).

It is divided into several subgenera, and those with species containing alkaloids of the plumerane types have been grouped together in Table 8. It is of chemotaxonomic importance that the alkaloids isolated so far from five of these six subgenera have the one absolute configuration only. The alkaloids of *Macrocarpa* are exclusively of the 7A-α type, and the other four, namely, *Nitida*, *Nobiles*, *Polyneura*, and *Quebrachines*, all have type β; *Pyricolla*, however, seems to be anomalous. The occurrence of the β-type in these four subgenera is of importance inasmuch as several structural types (4, 5, 6D, and 7D) occur together. From a consideration of the skeletal types present and their absolute configurations, one would be inclined to question the botanical subdivision of the

Table 8 The Genus Aspidosperma and its Subgenera. Distribution of Plumerane Alkaloids in the Various Species in Relation to Absolute Configuration

Subgenera, Species	Alkaloids Isolated	α	β	% Alkaloids of Known Absolute Configuration
1. *Macrocarpa*				
A. *duckei*	4	xxxx		100
A. *macrocarpon*	4	xxxx		100
A. *verbascifolium*	5	xxxxx		100
2. *Nitida*				
A. *discolor*	4		xxx	75
A. *eburneum*	2		x	50
A. *marcgravianum*	2		xx	100
A. *oblongum*	1		x	100
3. *Nobiles*				
A. *album*	10		xxx	33
A. *exalatum*	2			0
A. *fendleri*	6			0
A. *limae*	8		xxxxx	63
A. *megalocarpon*	1		x	100
A. *melanocalyx*	3			0
A. *neblinae*	10		x	10
A. *obscurinervum*	6		xxxxxx	100
A. *sanwithianum*	1		x	100
A. *spruceanum*	2			0
4. *Polyneura*				
A. *cuspa*	1		x	100
A. *cylindrocarpon*	15		xxxxxxxxxxxxxxx	100
A. *dispermum*	3		xxx	100
A. *peroba*	2			0
A. *polyneuron*	4		xxx	75
A. *sessiliflorum*	1			0
5. *Pyricolla*				
A. *gomezianum*	1			0
A. *multiflorum*	1			0
A. *olivaceum*	1		x	100
A. *populifolium*	7			0
A. *pyricollum*	2			0
A. *pyrifolium*	4	xx		50
A. *quirandy*	1			0
A. *refractum*	4	x		25
6. *Quebrachines*				
A. *chakensis*	3		xx	67
A. *quebracho-blanco*	11		xxxxx	42

Aspidosperma. On chemotaxonomic grounds, the following subdivision would suggest itself:

1 *Macrocarpa* (α, only 7A)
2 *Nitida* and *Nobiles* (β; 4, 5, 6D; exception 7D)
3 *Polyneura* and *Quebrachines* (β; 4, 5)
4 *Pyricolla* (α,β; 5, 6A)

Nitida and *Nobiles* are closely related botanically, and this is in agreement with chemotaxonomic evidence. However, *Polyneura* and *Quebrachines* are distinctly separated from one another. One might conclude from these results that in a comparison of chemical and botanical taxonomy the latter is superior, but such a conclusion would be too hasty; it is based exclusively on a study of a single class of alkaloid. It is easy to show that a better agreement with the botanical classification can be achieved if data from other classes of indole alkaloids occurring in *Aspidosperma* spp. are taken into account (see [H 14], [H 15]). Despite all this, the *Aspidosperma* spp. *A. exalatum* and *A. nublinae* can be classified in the subgenus *Nobiles* on the basis of their contents of alkaloids of the plumerane type; the botanical classification of these two species is not clear-cut from the literature.

7.2.3 Further Structural Observations

Among the various structural pecularities shown by the plumerane alkaloids (epoxide ring, presence of C-atom 22 on C-3,[8] oxidation number of certain carbon atoms) which can be used for chemotaxonomic evaluation, only the substitution pattern in the indolic nucleus would appear to be of some value. Data resulting from a study of these patterns are shown in Table 9. The alkaloids referred to—excluding those with unsubstituted aromatic nuclei—show five different patterns of substitution. No distinction is made between the substituents, which include $-OH$, $-OCH_3$, and O-alkyl (type 7D). It is remarkable that the methylenedioxy group, which is frequent in other types of alkaloids (e.g., isoquinoline alkaloids), does not occur here.

Apart from the tribe Carisseae, about 36% of all the alkaloids isolated have oxygen substituents on the benzene ring, the most important center being C-17. All tribes provide examples substituted at this position. From the table, similarities appear between two groups of plants, both belonging to the tribe Alstonieae: the genus *Aspidosperma* (except *Macrocarpa*) has the substitution

---

[8]So far it has not been possible to fit into the scheme, in a meaningful way, the substituent on C-3 consisting of carbon atom 22, which is derived from the terpenoid portion of the alkaloid.

Table 9 Pattern of Oxygen Substitution in the Plumerane Alkaloids

Apocynaceae	Number of Alkaloids Isolated	Percentage[a]	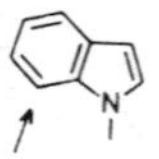 a	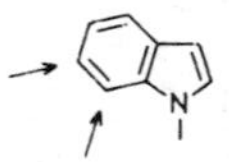 b	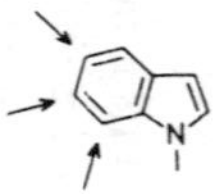 c	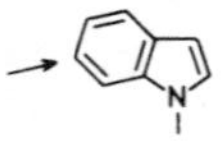 d	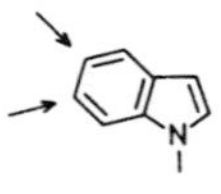 e
1. Carisseae							
Melodinus	28	7	1	—	—	1	—
Hunteria	9	0	—	—	—	—	—
Pleiocarpa	38	0	—	—	—	—	—
4. Tabernaemontaneae							
Rejoua	2	50	1	—	—	—	—
Ervatamia	1	0	—	—	—	—	—
Hazunta	2	100	—	—	—	1	1
Callichilia	6	50	3	—	—	—	—
Hedranthera	17	55	9	—	—	—	—
Tabernaemontana	14	21	3	—	—	—	—
Conopharyngia	4	25	1	—	—	—	—
Pandaca	14	0	—	—	—	—	—
Capuronetta	3	0	—	—	—	—	—
Tabernanthe	1	0	—	—	—	—	—
Stemmadenia	4	0	—	—	—	—	—
Crioceras	6	33	2	—	—	—	—
Voacanga	56	45	25	—	—	—	—

5. Alstonieae						
Craspidospermum	4	50	—	—	—	2
Gonioma	5	0	—	—	—	—
Alstonia	9	22	—	—	—	2
Aspidosperma						
1. Macrocarpa	13	0	—	—	—	—
2. Nitida	9	55	3	2	—	—
3. Nobiles	50	84	6	25	11	—
4. Polyneura	26	81	20	1	—	—
5. Pyricolla	21	67	9	5	—	—
6. Quebrachines	14	50	4	3	—	—
Rhazya	35	0	—	—	—	—
Amsonia	22	0	—	—	—	—
Vinca	40	25	—	—	—	10
Catharanthus	48	54	—	—	—	26
Haplophyton	5	100	3	—	2	—
6. Rauvolfieae						
Cabucala	3	67	—	—	—	2
Vallesia	14	64	9	—	—	—
Kopsia	10	10	1	—	—	—

[a]The percent figures give the proportion of alkaloids with nuclei a-e; the indole nucleus includes examples with a chromophore such as indoline. The percent figures add up to 100% if the unsubstituted representatives (not shown) are included.

patterns b and c as well as a; however, it does not have pattern d, which is significant in other genera. *Haplophyton*, which otherwise shows no special relationship with *Aspidosperma*, likewise forms an exception in this respect. The close relationship between *Catharanthus* and *Vinca* can be plainly seen from the table.

The substitution pattern e, which occurs in many indole alkaloids of the yohimbane, heteroyohimbane, and corynantheane types, is not typical for the plumerane alkaloids (see [H 14], [H 15]).

7.3 RESULTS

The most numerous group of indole alkaloids is formed by bases of the plumerane type, which is considered to include compounds related structurally and biogenetically to aspidospermidine (see Scheme 14). The occurrence of these bases is confined to the family Apocynaceae, within which they occur in species of a single subfamily, Plumieroideae. Their occurrence is thus quite specific, and a plant that contains plumerane alkaloids and has not been characterized botanically must thus belong to the Apocynaceae-Plumieroideae.

The subfamily Plumieroideae is divided botanically into seven tribes, four of which have so far been found to contain plumerane alkaloids. The latter include several structural types which are set out in Scheme 14, and it is apparent that the tribes of the Plumieroideae can be distinguished chemotaxonomically on the basis of the distinctive occurrence of these skeletal types. Some types occur in only one tribe, such as 6E in Tabernaemontaneae (Scheme 16) and 6C in Alstonieae.

The absolute configuration of alkaloids provides another significant factor which is of assistance in characterizing a tribe. It has been established that the biosynthesis of the plumerane alkaloids becomes more stereospecific with the increasing number of rings in the skeleton. While the fundamental types 4 and 5 with four or five rings occur in both α- and β- configurations, the same types occur in only one absolute configuration with more rings. (In this connection α and β refer to the configuration at center 5.) The fundamental types 6A, 6B, 6E, 7A, 7B, and 7C (see Scheme 14) thus have the α- configuration, while types 6D and 7D are exclusively β. Unfortunately a consideration of absolute configuration cannot include as many examples as one would wish, since many of the necessary data are lacking in the literature (only about 50% of the isolations recorded for alkaloids contain data that are useful in this respect). Nevertheless it has been possible on the basis of absolute configuration to further subdivide the genus *Aspidosperma* (see Table 8).

A chemotaxonomic study of compounds in relation to biogenetic pathways would appear promising. As many characteristics of a class of compounds as possible (e.g., skeletal type, absolute configurations, pattern of oxygen substitution, positions of C,C double bonds) should be used in a study of structural features. With this in mind we have studied the suitability in this regard of the oxygen substitution pattern on the indolic nucleus as well as the skeletal type and the absolute configuration. A sequence of products related to the alkaloidal precursors, which can be regarded as occupying a higher place in the biogenetic scheme, occurs in every tribe (see Schemes 15-18). As pointed out previously, wherever possible all the substances produced by the plant should be taken into consideration for a chemotaxonomic study. However, the great mass of data involved would present almost insurmountable chemical and analytical problems, apart from difficulties of retrieval from the literature, and for this reason such studies have mostly been restricted to one type of compound. In this book we have selected for study the plumerane alkaloids, which constitute only one type of indole base, and these in turn belong to a wider group of natural products, the alkaloids. The plumerane bases form only a small part of the total number of alkaloids (ca. 4%), but in spite of this they have provided the basis for chemotaxonomic deductions. Evidently their structures are sufficiently complex to be synthesized by only one plant family, but not so involved as to be restricted to one genus or species. They thus form an ideal class of substances for chemotaxonomy.

Structural Elucidation by Modern Methods: Villalstonine

In this section the methods that have been established for determining the structures of alkaloids are illustrated by reference to the structural elucidation of villastonine.

8.1 OCCURRENCE

The occurrence of this base is restricted to the genus *Alstonia*, which belongs to the family Apocynaceae. It was first isolated in 1934 from the bark of *A. macrophylla* Wall., *A. somersetensis* F. M. Bailey, and *A. villosa* Blume [S 40]. (The name of the alkaloid is derived from its occurrence in the latter plant.) Villastonine was also found later in *A. muelleriana* [N 3], *A. glabriflora*, and *A. spectabilis* [H 19]. In all the above-mentioned plants it does not occur alone, being accompanied by other alkaloids from which it can be separated by chromatographic methods.

The alkaloid content of a plant is usually not the same in all parts (e.g., roots, heartwood, bark, flowers, fruit, stalks). In many cases alkaloids are even restricted to a single part of a plant; for example, aphelandrine occurs only in the roots of *Aphelandra squarossa* var. Roezii. One can moreover often observe very considerable variations in the composition of a mixture of alkaloids when different parts of an alkaloid-containing plant are examined. Another factor that can influence the alkaloid content is the season of the year in which the plant has been collected (e.g., during or after the time of flowering); the type of soil can also make a difference. It is thus useful to have as much detail as possible at one's disposal before making a study of a plant extract. The villalstonine described here was obtained from the trunk bark of *Alstonia macrophylla*.

8.2 PHYSICAL PROPERTIES

Before a chemical study of any compound can be made, it must be characterized, that is, its specific properties such as melting point, optical rotation, and spectroscopic data must be determined. For this purpose a fundamental assumption is that the substance is pure and homogeneous. Usually tests for purity are carried out involving the use of chromatographic methods such as TLC, together with recrystallization and determination of melting point. The purity is checked by elemental analyses for carbon and hydrogen.

Villastonine (8/1) melts at 235-270°C with decomposition, and has $[\alpha]_D^{26°}$ = +79° ± 5° (c = 1.6; chloroform). Its elemental composition was determined as $C_{41}H_{48}N_4O_4$ (M = 660.864) by mass spectroscopy (M^+ = 660.3676).[1]

Nowadays the molecular weight of an alkaloid is frequently determined by mass spectroscopy alone, using electron impact ionization at 70 eV. However, this method involves certain possibilities of error that one should be aware of when the molecular weight of a structurally unknown compound is determined in this way. The sample must be vaporized in order to obtain the mass spectrum, and the temperature of vaporization depends on several factors, including the molecular size and the number and kind of functional groups present. During vaporization the possibility of thermal reactions taking place cannot be excluded. These could lead to rearrangements as well as to the loss of small or large neutral fragments. The products that may be formed in this way, after ionization, will give a false molecular weight because of dehydration (M-18), decarboxylation (M-44), decarbonylation (M-28), and so on. Other processes, such as thermal transmethylation, are known which can result in the molecular weight recorded by the mass spectrometer being too high. The spectrum may show (M + 14) and (M + 28) peaks with the same intensity as the molecular ion which is recorded simultaneously. More detailed descriptions of these and other thermal reactions which take place in the mass spectrometer are given in reviews ([B 22], [V 1]; see also Section 8.4). With improvements in the techniques of insertion and with modifications in the ionization procedures the danger of thermal reactions in mass spectrometry is diminished but not completely removed. The field desorption method of ionization is especially reliable in this respect [R 9].

It is not always easy to determine whether the ion of highest mass recorded by the mass spectrometer arises from the original substance, or whether it has undergone thermal change. For this reason it is advisable to check the molecular weight by other means, such as osmometric methods or combustion analysis. The latter also gives a good idea of the purity of the substance—thus hydro salts

---

[1]The difference in mass results from different relative values; the mass spectrosopic values are for pure isotopes.

of alkaloids decompose thermally in the mass spectrometer into the free base and the acid. The mass spectrum may hardly be distinguished from that of the pure alkaloid, but combustion analysis gives information on the salt.

Villalstonine shows two pK_a values (in 80% methylcellosolve): 5.39 and 6.98, from which it follows that two of the nitrogen atoms are basic and have different environments. The other two nitrogens must be neutral, provided that two transition points do not happen to coincide. The dibasic nature of the alkaloid is also evident from its salt-forming properties. Villalstonine dihydrochloride (8/2), which is precipitated on addition of an ether solution of villalstonine to dry ethereal hydrogen chloride, is a crystalline compound. N,N-Dimethylvillalstonine diiodide (8/3, villalstonine dimethiodide) cannot be obtained pure under ordinary reaction conditions, such as the addition of methyl iodide at room temperature to a solution of the base in benzene containing a little methanol and allowing to stand for a few hours. However, the quaternization takes place smoothly if sodium carbonate is added to the reaction mixture to neutralize the hydrogen iodide formed from the methyl iodide and the methanol-containing solvent, and then the reaction mixture is boiled for a few hours under reflux.

The presence of 1.1 OCH_3 and 2.1 NCH_3 groups in villalstonine was established by Zeisel determination.

8.3 SPECTRA AND FUNCTIONAL GROUPS

Two intense bands appear at 1754 and 1730 cm^{-1} in the ester carbonyl region of of the infrared (IR) spectrum of villalstonine in chloroform (see Figure 1).

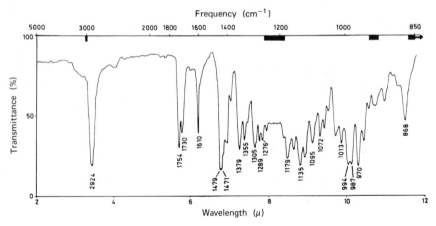

Figure 1 Infrared spectrum of villalstonine in CHCl$_3$ (0.2 mm microcell; the sections marked in black show the regions where solvent absorption occurs).

Another strong absorption appears in this region at 1610 cm^{-1}. These three bands will be referred to later. There are no absorptions corresponding to OH or NH groups, and in agreement with this, villalstonine cannot be acylated under normal conditions (pyridine/acetic anhydride, 1/1; 45°). It does not appear possible to make an unambiguous analysis of the remaining regions of the spectrum.

As one might expect from the large number of hydrogen atoms, the ^{1}H-nuclear magnetic resonance (NMR) spectrum (Figure 2)[2] is quite complex. A multiplet that integrates for seven aromatic protons appears in the region between 6.5 and 7.7 ppm, and a further aromatic proton forms a doublet ($J =$ 8 Hz) at 6.16 ppm (C-12 H). An incompletely resolved doublet of doublets arising from a vinyl proton (at C-19) is centered at 5.37 ppm, and a doublet can also be recognized at 4.43 ppm (J = 4 Hz) (from the C-16 H). Methyl proton singlets occur at 3.58 and 3.64 ppm (COOCH$_3$ and indolic NCH$_3$), 2.28 ppm (aliphatic NCH$_3$), and 1.23 ppm (C-19'CH$_3$), and a methyl doublet with fine structure ($J \approx$ 8 Hz) occurs at 1.55 ppm (C-19 CH$_3$).

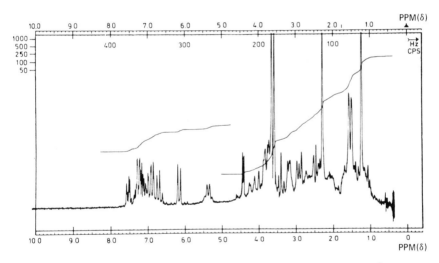

Figure 2 ^{1}H NMR spectrum of villastonine (8/1) in CHCl$_3$ at 100 MHz[2].

The UV spectrum of villalstonine (in 95.5% ethanol) is given in Figure 3. An analysis reveals that the alkaloid does not have any known simple chromophore. The mass spectrum of villastonine will be discussed later.

A few simple reactions were carried out in order to clarify the nature of the functional groups. When villastonine is treated with lithium aluminum hydride

---

[2]All NMR spectra are at 100 MHz; CDCl$_3$ was used as solvent unless otherwise stated, and tetramethylsilane (TMS) as internal standard.

(log ε) + Z

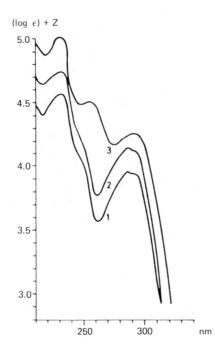

Figure 3 UV spectra in 95.5% ethanol of villalstonine (**8/1**) and villamine (**8/16**), and the combined spectra of macroline (**8/18**) and 2,7-dihydropleiocarpamine (**8/15**).
Curve 1: Villastonine (M = 660), c = 4.77 × 10^{-5}, Z = 0. Maxima: 231 (4.57), 286 (3.96) Minima: 216 (4.40), 261 (3.57) Shoulders: 250 (4.00), 294 (3.93).
Curve 2: Villamine (M = 660), c = 4.62 × 10^{-5}, Z = 0.2. Maxima: 231 (4.56), 286 (3.97) Minima: 216 (4.46), 260 (3.60) Shoulders: 250 (3.96), 294 (3.93).
Curve 3: Combined sepctra of vacroline + 2.7-dihydropleiocarpamine, Z = 0.4. maxima: 230 (4.61), 252 (4.11), 292 (3.85) Minima: 218 (4.46), 247 (4.08), 274 (3.77).

in tetrahydrofuran (8 hr/20°C), the alcohol villalstoninol (**8/4**, M = 632) is formed, which shows some characteristic differences as compared to villalstonine. As a result of the loss of 28 mass units (amu) from the molecular mass, a methoxycarbonyl group must have been reduced to a primary alcohol:

$$R\text{-}COOCH_3 \ (R + 59 \ amu) \longrightarrow R\text{-}CH_2OH \ (R + 31 \ amu); loss = 28 \ amu$$

In conformity with this conclusion, compound **8/4** can be acetylated (pyridine/acetic anhydride) to give a monacetyl derivative. The UV spectra of **8/1** and **8/4** show no significant difference, and thus the methoxycarbonyl group in **8/1** is not conjugated with the rest of the molecule. It is noteworthy that both ester carbonyl bands are lacking in the IR spectrum (CCl_4) of **8/4**. Intense absorptions occur only at 3030 (OH) and 1662 cm^{-1}, from which it must be concluded that the carbonyl bands at 1754 and 1730 cm^{-1} in the IR spectrum of villalstonine originate from a single ester carbonyl group, otherwise the mass difference of 28 amu between villalstonine and villalstoninol cannot be explained. This conclusion is supported by the ^{1}H-NMR spectrum of **8/4**, which shows three methyl singlets (at 3.70, 2.36, and 1.31 ppm) as compared to four for villalstonine.

Villalstonine can be reduced to dihydrovillalstonine (8/5) by hydrogenation over a palladium catalyst, a process that involves reduction of an isolated double bond (8/5 has almost the same UV spectrum as 8/1, but the molecular weight has increased by 2 amu). In the ^{1}H-NMR spectrum of 8/5, the quartet at 5.37 ppm and the doublet at 1.55 ppm originally present are lacking, from which it may be deduced that villalstonine has a trisubstituted ethylidine group (C=CH–CH$_3$), and moreover the quartet at 5.37 ppm is established as a standard for the integration of one proton (1H) in the spectrum of 8/1.

Two of the four nitrogen atoms present in villalstonine (8/1) are tertiary and basic (see Section 8.2). On treatment of 8/1 with methyl iodide containing a little methanol and anhydrous sodium carbonate, villalstonine dimethiodide (8/3) is formed after refluxing for 24 hours. The elemental composition of this salt was determined by combustion analysis as $C_{43}H_{54}I_2N_4O_4 \cdot 3H_2O$,[3] and the molecular ion on mass spectrometry was found to be 660, the same as for 8/1. Thus the dinorbase villalstonine (8/1) is formed in the mass spectrometer by thermal decomposition. If one of the nitrogen atoms were secondary or primary, the molecular weight of the dinorbase, apart from the results of the combustion analysis, would be higher by 14 or 28 amu, corresponding to the reaction N–H $\longrightarrow$ N–CH$_2$–H.

While salts of organic bases and acids undergo ready thermal decomposition into the corresponding base and acid[4] which on electron impact are ionized independently of one another in the mass spectrometer, the removal of a proton from the thermally stable tetraalkyl ammonium compound does not take place, since in the latter the substituent attached to the nitrogen is an alkyl residue. The vaporization of a salt is achieved only by its conversion into neutral fragments.

---

[3]Perchlorate (ClO$_4^-$), picrate (C$_6$H$_2$N$_3$O$_7^-$), and iodide (I$^-$) are especially suitable as anions for tetraalkyl ammonium compounds. They give quaternary nitrogen salts that can usually be recrystallized from acetone/water. The water content of crystallization is low (generally between zero and 3H$_2$O). Other salts such as bromide, chloride, fluoride, and sulfate crystallize much less easily, and often contain more water in their crystalline lattice. Usually they will crystallize only from higher alcohols (propanol, isopropanol, etc.) or from lower alcohols plus ether. Precipitation with saturated sodium perchlorate, picrate, or iodide solution from an aqueous solution of the alkaloid chloride or fluoride has proved a useful method for the interconversion of anions. Conversely alkaloid picrates or iodides (in acetone/water = 1/1) can be converted into the corresponding chloride or fluoride on an ion-exchange column, using a chloride or fluoride ion-exchange resin. The commercially available exchange resins saturated with chloride ion can be converted into the fluoride form by washing with an excess of saturated potassium fluoride solution.

[4]In adduct ion mass spectroscopy the "molecular ion" of the salt is observed in certain cases (e.g., papaverine hydrochloride). This is presumably not the true molecular ion, but an addition product formed from the base and the acid. This type of addition product is encountered relatively frequently in adduct ion mass spectroscopy.

The thermal conversion of a salt into neutral molecules takes place according to well-defined rules which allow conclusions to be drawn concerning its structure. Tetraalkylammonium compounds with the general formula 8/6 (X = halogen) can be converted into neutral molecules in three ways.

1 **DESALKYLATION**   The anion attacks an alkyl group on the quaternary nitrogen atom with formation of a tertiary amine (norbase) and an alkyl halide.

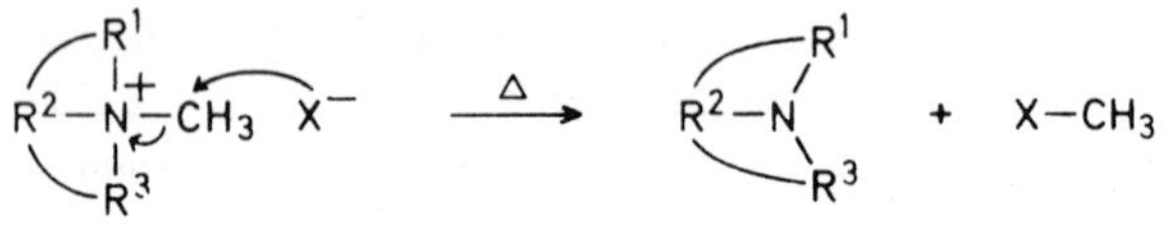

8/6

This mode of degradation is preferred with iodide ions, but takes place seldom with fluoride ions. Strychnine methiodide (8/7, molecular weight 476) serves as an example for this kind of degradation. In the mass spectrum of the mixture formed by thermal decomposition, the two molecular ions appear at 334 (strychnine) and 142 (methyl iodide), and each gives a spectrum identical with the corresponding pure compound.

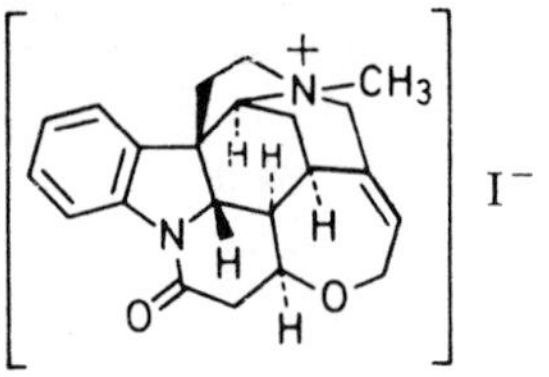

8/7

2 **THERMAL HOFMANN DEGRADATION**   The anion attacks a hydrogen atom attached to a β-position with regard to the quaternary nitrogen atom (see 8/8), forming a tertiary amine (methine base 8/9) and hydrogen halide.

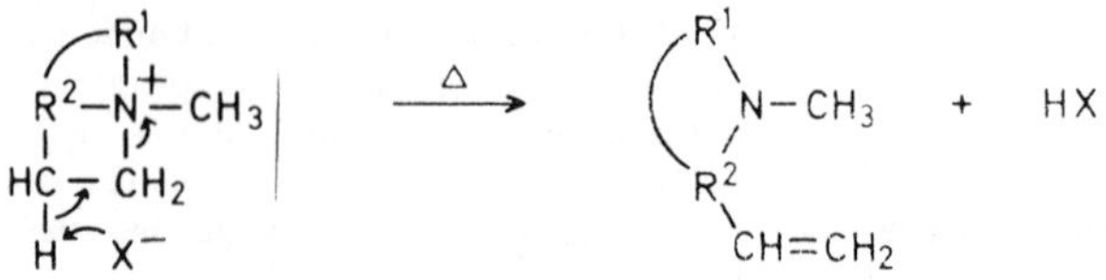

8/8   8/9

In a thermal dealkylation of tetraalkylammonium compounds of the general type 8/6 there is only one possibility for demethylation. On the other hand, under favorable conditions (at least three hydrogen atoms attached to different carbon atoms located β to the quaternary nitrogen), three different Hofmann degradations can in principle occur, and this number can be increased by reactions involving vinylogous or ethylogous processes. The thermal Hofmann degradation is given principally by alkyl fluorides and chlorides, and much less frequently by bromides and iodides. In all cases so far studied, only one main degradation product is formed. Thus the quaternary compound 8/10 (ϵ_2-dihydromavacurine methochloride) is degraded thermally in the mass spectrometer into the α-methyleneindoline base 8/11. A side reaction that is often observed consists in the dimerization of the Hofmann base, giving rise to intense $2M^+$ peaks; a diene addition is probably responsible for the appearance of these products.

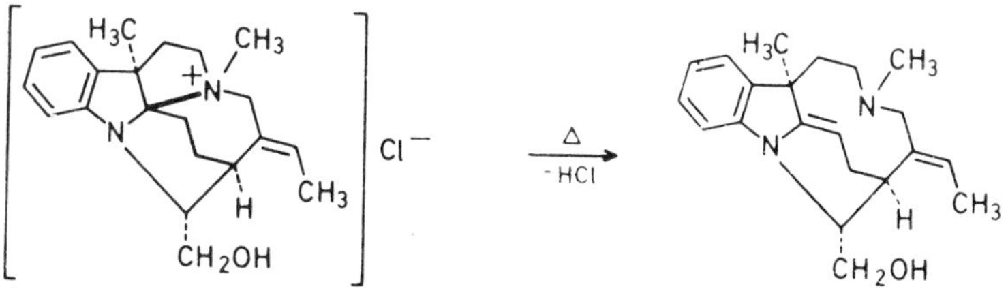

8/10   8/11

3 SUBSTITUTION REACTION   The anion attacks another carbon substituent on the quaternary nitrogen, forming a tertiary base that contains the anion as a substituent.

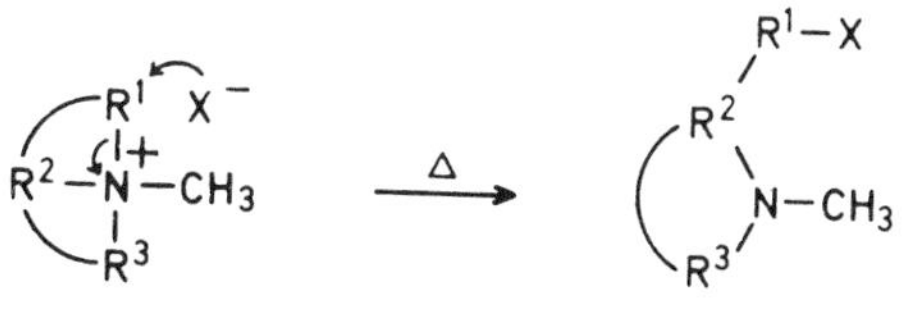

8/6

The desalkylation and substitution reactions differ from one another only in the point of attack; the latter is a special case of the first reaction.

The thermal substitution reaction seems to depend largely on the structure of the cation and so far has been seldom observed. In a certain sense this reaction involves an isomerization, since the molecular weight of the salt is the same as that of the pyrolysis product. So far the substitution reaction has been observed only with metho salts of pleiocarpamine and a few derivatives thereof. The structure of the compound 8/13 has been determined chemically [V 1].

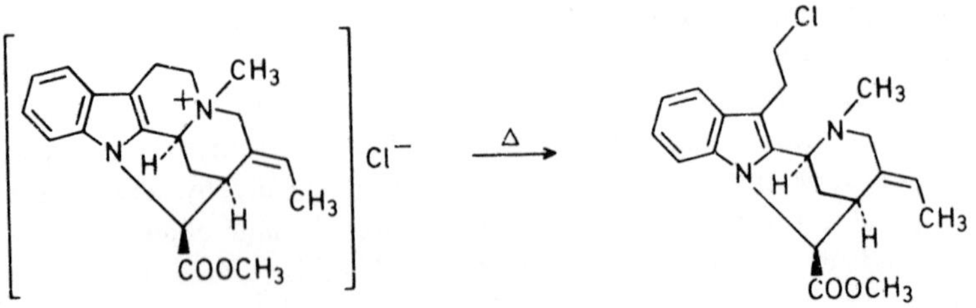

8/12   8/13

It may be noted that cations of tetraalkylammonium salts can be recorded by alkali ion adduct mass spectroscopy.

The results obtained so far in the structural elucidation of villalstonine can be summarized as follows:

1 Groups definitely present in villalstonine (8/1):

1   $\diagdown$C=C$\diagup$ (H, CH$_3$)   (from NMR, catalytic hydrogenation)

1   $-COOCH_3$ (from NMR, LiAlH$_4$ reduction; both are stronger arguments than IR, which indicates two $COOCH_3$ groups)

8   aromatic H   (from NMR)

2   basic $\diagdown$N$-$ (from dimethiodide 8/3)

2   $\diagdown$N$-CH_3$ (from NMR, N$-$CH$_3$ determination; one of them is presumably attached to a basic N from NMR)

1   $-\overset{|}{\underset{|}{C}}-CH_3$ (from NMR)

2 Groups definitely not present in 8/1:

$-OH$ and $\diagdown$NH groups   (from acetylation and methylation experiments) other $\diagdown \diagup$ C=O   (from IR)

It is evident from this list that, so far, the methods of investigation have given few results. They have not provided a real insight into the molecular structure, or even an indication of the type of alkaloid. The next step is to try to achieve this by chemical reactions. Suitable points at which to apply these reactions are provided by functional groups such as amide, ester, and ether, which may undergo cleavage or rearrangement with the appropriate reagents. Another starting point may be provided by the mass spectrum in which signals indicating a relationship to known alkaloids or known structural elements may be observed. In that case it may be worthwhile applying reactions that are characteristic for those compounds or structural elements.

In the case of villalstonine there are several leads of this sort that can be followed up, but it must be emphasized that they are not conclusive and are merely indications suggesting the direction that investigations might take. Leads of this kind are:

1   The above-mentioned double band in the carbonyl region of the IR spectrum produced by the methoxycarbonyl group is a feature that is seldom observed. It might be supposed that another compound showing this phenomenon could serve as a model for villalstonine (**8/1**), and that the same or a similar group might be present in both compounds. The indole alkaloid pleiocarpamine (**8/14**) also shows a double band in its IR spectrum corresponding to a methoxycarbonyl group.[5] It is thus possible that pleiocarpamine is a derivative of villalstonine or forms an essential structural element of it.

2   Before the structural elucidation work on villalstonine was commenced it was known that indole alkaloids had been isolated from other plants of the genus *Alstonia*, which belongs to the Apocynaceae, a family known to contain indole alkaloids (see Section 7). The assumption is thus justified that villalstonine is also an indole alkaloid. The IR spectrum of villalstonine provides a further clue. The band at 1610 cm^{-1} corresponds to an absorption observed with indoline (dihydroindole) chromophores. If villalstonine has this chromophore,[6] it should give the color reaction common to indoline alkaloids, which

---

[5] The IR spectrum of pleiocarpamine (**8/14**) shows the double band at 1736 and 1770 cm^{-1} (CCl$_4$); at 1736 and 1767 cm^{-1} (CS$_2$); at 1724 cm^{-1} only (KBr); and at 1730 cm^{-1} only (Nujol).

[6] The UV spectrum should of course show the indoline chromophore. However, the spectrum is complex as already mentioned. There are at least two overlapping chromophores, so that an evaluation at the present stage is not possible.

is produced by the cerium(IV) sulfate reagent. However, in the case of villalstonine the reaction is negative, an observation in direct contradiction to the IR spectroscopic evidence.

A possible cause of failure of the color reaction here is the acid present in the reagent, which consists of a 4% aqueous solution of $Ce(SO_4)_2$ in $2N$ H_2SO_4.[7] The acid could be producing a deep-seated molecular rearrangement. This conjecture, although vague, could provide a starting point for further structural investigations on villalstonine (8/1).

One can arrive at the same conclusion from another point of view, and thus make a decision concerning the subsequent course of action. As already mentioned, villalstonine contains four oxygen atoms, two of which are in a methoxycarbonyl group. So far there is only negative evidence concerning the other two oxygen atoms, which, according to spectroscopic evidence, cannot be present in hydroxy or carbonyl groups. Other oxygen-containing functional groups which often occur in natural products are ethers and acetals. The latter can be hydrolyzed with strong acids, and from this consideration it is also worthwhile to study the influence of acids on villalstonine (8/1).

8.4 ACID-CATALYZED CLEAVAGE REACTIONS OF VILLALSTONINE

If villalstonine is treated with 70% perchloric acid for 45 minutes at 20°C, a purple-colored solution is obtained. After neutralization and extraction, this gives a product that is purified by chromatography on alumina to afford pleiocarpamine (8/14, $C_{20}H_{22}N_2O_2$, $M = 322$) in 40% yield. The pleiocarpamine obtained in this manner shows the same properties (UV, IR, NMR, mass spectroscopy, $[\alpha]_D$, chromatographic behavior, color reactions, and m.p.) as the natural alkaloid from the apocynaceous plant *Pleiocarpa mutica* Benth. whose structure had already been determined. It was not possible to isolate any other pure compound from the perchloric acid hydrolyzate, although the other

---

[7]Other color reagents frequently used in alkaloid chemistry:

Potassium iodoplatinate reagent: solution (1): 1 g $PtCl_4 \circ 2H_2O$ + 6 ml H_2O + 20 ml $1N$ HCl; solution (2): 9 g KI + 90 ml H_2O. Mix in the proportion of 1 ml (1) + 9 ml (2) + 20 ml H_2O [S 55].

Dragendoff's reagent: 8 g $Bi(NO_3)_3 \cdot 5H_2O$ in 20 ml HNO_3 (specific gravity 1.18) + 27.2 g KI in 50 ml H_2O. After mixing, KNO_3 crystallizes out; the solution is decanted and diluted to 100 ml [C 6].

Mayer's reagent: 1.36 g $HgCl_2$ in 60 ml H_2O + 5 g KI in 10 ml H_2O; dilute to 100 ml with H_2O.

8/14, (+)-*Pleiocarpamine*

products were examined with great care. Most of the original villalstonine was converted into resinous products.

Villalstonine cannot be just a dimer of pleiocarpamine, since its molecular weight is not simply double that of pleiocarpamine. There must thus be another unit of slightly higher molecular weight joined to pleiocarpamine, which is formed under acid conditions but destroyed thereby. This suggests a study of the cleavage reaction under carefully controlled acid conditions, with the object of trying to isolate the other compound. Acetic acid and hydrochloric acid both proved too mild, and the experiment resulted in recovery of the starting material. Sulfuric acid and phosphoric acid led to the isolation of pleiocarpamine, but in poorer yield than with perchloric acid. The product formed is clearly very sensitive, and this suggests the possibility of reducing it as soon as it is liberated. Thus experiments were tried in which villalstonine was treated with hydrochloric acid in the presence of tin, zinc, or tin(II) chloride under reducing conditions. In all these cases the second compound was not isolated, but only 2,7-dihydropleiocarpamine (8/15, $M = 324$). It was established in a control experiment that pleiocarpamine is converted into 8/15 under the same reductive conditions, and this experiment thus gave no further information about the structure of villalstonine.

After a number of fruitless hydrolytic experiments using acid conditions, a breakthrough was finally achieved with a mixture of trifluoroacetic acid and trifluoroacetic anhydride. This mixture was allowed to react with crystalline villalstonine dihydrochloride tetrahydrate ($C_{41}H_{48}N_4O_4 \cdot 2HCl \cdot 4H_2O$) for 5 minutes at room temperature; then the reagent was removed in a current of nitrogen and the residue was basified with ammonia and worked up. In addition to a little starting material, two isomers of villalstonine were isolated. The main product was called villamine (8/16, yield 40.5%), and the minor product villoine (8/17, yield 25.4%). Under these reaction conditions no pleiocarpamine (8/14) was formed.

The following points may be noted as far as these two new substances are concerned:

1 The above-mentioned double band due to the methoxycarbonyl group appears in the IR spectra of both compounds: in **8/16** at 1762 and 1737 cm^{-1} (CCl$_4$), and in **8/17** at 1754 and 1731 cm^{-1} (CDCl$_3$). It may be deduced that both compounds contain pleiocarpamine as a building block.[8]

2 Both villamine and villoine are converted back into villalstonine with dilute hydrochloric acid.

3 The mass spectrum of villamine (**8/16**), shown in Figure 4, provides a very important clue. It differs in quite a characteristic way from that of villalstonine (**8/1**, Figure 5), but shows a certain similarity to that of pleiocarpamine (**8/14**, Figure 6).

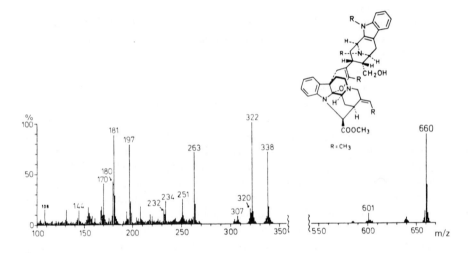

Figure 4 Mass spectrum of villamine (**8/16**).

The relationship between these compounds is expressed by the fact that the mass spectrum of pleiocarpamine forms part of the mass spectrum of villamine (**8/16**). If one subtracts the former from the latter spectrum, and if one neglects the signals at higher mass (zbove m/z 400), one obtains a new spectrum of great interest since it could represent the mass spectrum of the second fragment we had been seeking to isolate. The basis of this assumption depends on simple arithmetic. The molecular weight of the fragment pleiocarpamine is 322; the ion at m/z 338 is presumably the molecular ion of the other fragment. The two

[8] In addition to pleiocarpamine, a double band in the IR spectrum is also shown by 2,7-dihydropleiocarpamine and 16-*epi*pleiocarpamine for the ester carbonyl group.

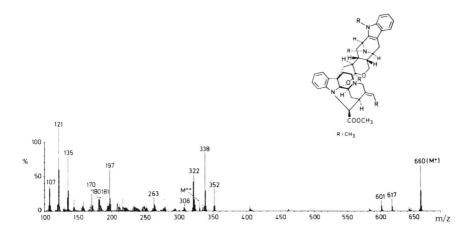

Figure 5 Mass spectrum of villalstonine (8/1).

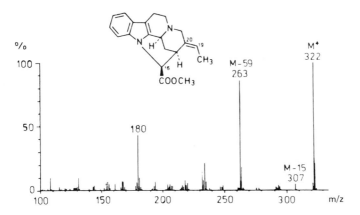

Figure 6 Mass spectrum of pleicarpamine (8/14).

figures add up to 660, the molecular weight of villalstonine (8/1), villamine (8/16), and villoine (8/17). This argument is, of course, speculative and provides no more than a hint, but it is worth looking into, especially as all previous attempts to obtain the other fragment from villalstonine were unsuccessful. We should therefore test this hypothesis before discarding it.

Let us consider briefly the origin of the mass spectrum of an organic compound that does not volatilize easily. Before the molecules of the test sample can ionize, the sample must vaporize, and during this process thermal reactions can take place, as we have already seen in some detail. Any fragments formed in such reactions are ionized independently and give overlapping mass spectra. If a thermal reaction were to take place during the vaporization of villamine (8/16), the superposition of the mass spectra of pleiocarpamine (8/14) and the other fragment in the mass spectrum of villamine (8/16) would be explained, but not the appearance of signals with mass greater than m/z 400. A second possibility must therefore be kept in mind, that is, that villamine (8/16) gives the signals shown in Figure 4 by straightforward mass spectroscopic means. In order to test its thermal stability, villamine (8/16) was vacuum distilled.[9]

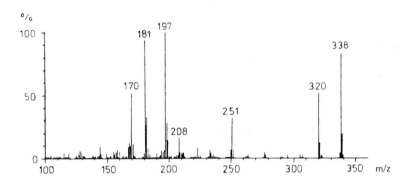

Figure 7 Mass spectrum of macroline (8/18).

The distillate afforded pleiocarpamine (8/14, $C_{20}H_{22}N_2O_2$, M = 322) as well as another compound which has been named macroline (8/18, M = 338). Evidently villamine decomposes at least partially in the mass spectrometer on

[9]Distillations of this kind are preferably carried out in a small flask which consists of a glass tube blown out into a small bulb at one end. A solution of the test substance is introduced directly into the vessel through a long funnel consisting of a drawn-out test tube. The solvent is removed by oscillating the bulb in warm water under vacuum so as to produce as thin a film as possible on the walls of the vessel. The actual pyrolysis is carried out by dipping the bulb for a short time in a heated metal bath with the end of the tube connected to a vacuum pump. The substance forms a light-colored lacquer inside the tube above the metal bath.

heating. There is a surprisingly good correspondence between the mass spectrum of pure macroline (Figure 7) and that obtained by the subtraction method.

A solution to part of the problem involved in the structural elucidation of villalstonine has thus been found. The following procedure suggests itself for tackling the rest of the problem:

1   Elucidation of the structure of the fragment macroline (8/18)

2   Derivation of the structure of villamine (8/16)

3   Formulation of the structure of villalstonine (8/1) itself

8.5   ELUCIDATION OF THE STRUCTURE OF THE FRAGMENT MACROLINE

(+)-Macroline (8/18, $C_{21}H_{26}N_2O_2$, M = 338) crystallizes from methanol/ether in colorless needles which melt with decomposition between 211 and 213°C. The specific rotation of macroline is $[\alpha]_D$ = +19° ± 5° (c = 0.406; methanol), and its UV spectrum (Figure 8) is the same as that of an N-alkylated indole, with λ_{max} 231 nm (log ϵ = 4.56), 282 nm (3.67), and 288 nm (3.66), and λ_{min} 254 nm (3.12) and 286 nm (3.64); in ethanol. This assignment was made by comparing the spectra from the literature with that of the "unknown" macroline. The IR spectrum in CCl_4 (Figure 9) shows the presence of an hydroxy group (weak band at 3205 cm^{-1}) and an α,β-unsaturated ketone (1681 and 1623 cm^{-1}). The band due to the C,C double bond at 1623 cm^{-1} is quite weak.

These results from the UV and IR spectra are confirmed and extended by the ^{1}H-NMR spectrum (60 MHz, CDCl$_3$, tetramethylsilane as internal standard). In the region between 6.8 and 7.8 ppm a multiplet corresponding to four aromatic protons appears. The N-alkyl group on the indole residue is a methyl, as shown by a singlet at 3.60 ppm (for villalstonine, 3.58 or 3.64 ppm). The α,β-unsaturated ketone recognizable from the IR spectrum is further characterized in the NMR spectrum. The C,C double bond proves to be in a vinylidine group—it gives a singlet at 6.18 ppm (1H) and a singlet with a small coupling ($J \approx 1.5$ Hz) at 5.96 ppm, which is produced by a proton in an allylic position. The splitting can be removed by irradiation of the allylic proton at a frequency of 2337 Hz. A methyl group, which gives a singlet at 2.26 ppm, is attached to the CO group on the other side of the α,β-unsaturated ketone. Since the nature of both oxygen atoms in macroline is now known (OH and C=O), a second keto group cannot be present, and the NMR assignments are unambiguous. Another signal at 2.38

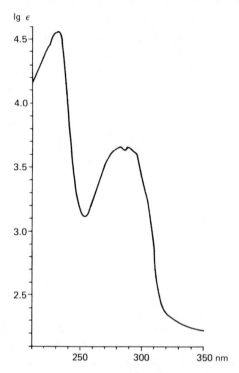

Figure 8 UV spectrum macorline (8/18) in 95.5% ethanol, $c = 4.47 \times 10^{-5}$. Maxima: 231 (4.56) 282 (3.67) 288 (3.66) Minima: 254 (3.12) 286 (3.64)

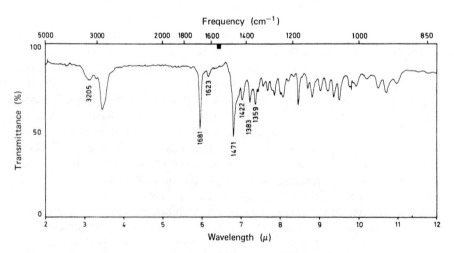

Figure 9 IR spectrum of macroline (8/18) in CCl_4 (0.2 mm microcell; the section marked in black shows the region where solvent absorption occurs.

128

ppm is easily assigned, being evidently due to a methyl group on a second basic nitrogen atom (for villalstonine, 2.28 ppm).

The following partial structure for macroline can be deduced from its UV, IR, and NMR spectra:

Partial structure I for macroline based on its UV, IR, and NMR spectra.

The actual nature of the ring system is not known. The assignments so far made on purely spectroscopic grounds must be confirmed by other means. The presence of the carbonyl group is shown by the reduction of macroline (8/18) with lithium aluminum hydride in tetrahydrofuran, which gives macrolinol (8/19, $C_{21}H_{28}N_2O_2$, $M = 340$) as the main product. This substance shows no essential difference in its UV spectrum to that of the starting material 8/18. In other words, the CO group in 8/18 is not conjugated with the chromophore. In the IR spectrum ($CHCl_3$) there is no band corresponding to a keto group, but very strong absorptions appear in the OH region. As before, the band due to the C,C double bond is weak, but the absorption due to the vinyl double bond at 913 cm^{-1} is easily recognized. The reduction of the CO group to the corresponding alcohol produces a considerable difference in the NMR spectrum. The signals due to the two vinyl protons undergo an upfield shift of about 1 ppm (to 5.20 and 5.05 ppm, both singlets), and in consequence lie in a region typical of the vinyl protons of allyl alcohols.[10] The methyl group adjacent to the CO group in 8/18 appears in the ^{1}H-NMR spectrum of 8/19 as a doublet at 1.15 ppm ($J \approx 6.5$ Hz).

Macrolinol (8/19) can be converted with acetic anhydride/pyridine into an O,O-diacetyl derivative (8/20; $M = 424$; IR: 1727 cm^{-1}, ester carbonyl), in accord with the presence of two hydroxyl groups. At the same time this shows that macroline (8/18) has no secondary nitrogen atom. (A primary N function

[10]From the tables of Silverstein and Bassler [S 43] the following chemical shifts for the vinyl protons are expected for 8/18 and 8/19, respectively, in ppm:

has already been excluded because of the two NCH_3 groups.) The monoacetylation of macroline does not take place smoothly, but one can obtain O-acetylmacroline (8/21) in good yield by pyrolysis of O-acetylvillamine (8/22), which can be prepared in the usual way from villamine (8/16).

The presence of the C,C double bond is confirmed chemically by catalytic hydrogenation (PtO_2 / H_2 / ethanol) of macroline (8/18) to 19,20-dihydromacroline (8/23, $M = 340$). From the data obtained so far, the partial structure II can be written:

Partial structure II for macroline (8/18) based on all the chemical and spectroscopic data, except the mass spectra.

An analysis of the mass spectra proved especially useful in clarifying the nature of the ring nucleus and the way in which the structural features shown in II are joined together. If one compares the main peaks in the mass spectrum of macroline (8/18, Figure 7) with those of its derivatives (see Table 10) a number of deductions which are important for the structural analysis can be made. We shall examine in particular the following six characteristic fragments, concentrating on the signals from macroline. Those of its derivatives are given for comparison.

1   The ions shown in Table 10 all contain an aromatic nucleus, as shown by the shift of these signals by 4 amu in the spectrum of d_4-macroline (8/24). The latter compound was prepared from villalstonine (8/1) by isomerization to villamine (8/16) under deuterating conditions $[(CF_3CO)_2O/CF_3COOD]$ followed by pyrolysis. It may be further deduced from this spectrum that the keto group in macroline (8/18) is no longer present in villamine (8/16).[11] This point will be taken up again later.

2   All the ions have at least one nitrogen atom, and the ions of mass 338 (M^+), 320, and 197 contain two from high-resolution (HR) data. As noted previously, macroline has two nitrogen atoms, of which one is part of the indole

---

[11] By acid catalysis aromatic protons, as well as acidic protons in α-positions to carbonyl, nitrile, and nitro groups, can be exchanged for deuterium. Protons attached directly to heteroatoms (OH, NH, etc.) can also be exchanged for deuterium, but under normal conditions, unless protons are carefully excluded, back exchange takes place so rapidly that usually only the original OH or NH is recorded.

Table 10   Characteristic Peaks in the Mass Spectra of Macroline (8/18) and Some of Its Derivatives

Macroline (8/18)	338 (M^+) HR $C_{21}H_{26}N_2O_2$	320 $C_{21}H_{24}N_2O$	251 (g) $C_{17}H_{17}NO$	208 (h) $C_{15}H_{14}N$	197 (a) $C_{13}H_{13}N_2$	181 (i) $C_{13}H_{11}N$	170 (d) $C_{12}H_{12}N$
m^*	• ⟶	⟶	• ⟶	⟶			
Macrolinol (8/19)	340	322	253	208	197	181	170
20,21-Dihydromacroline (8/23)	340	322	253	210	197	181	170
O-Acetylmacroline (8/21)	380	320	251	208	197	181	170
d_4-Macroline (8/24)	342	324	255	212	201	185	174
Macroline · CD$_3$I (8/25)	341 +	323 +	251	208	200 +	181	170
	338	320			197		

HR – High-resolution mass spectrometry.

131

chromophore and is thus weakly basic, while the other has definite basic properties. By treating macroline with methyl iodide, the quaternary salt macroline methiodide is obtained. If trideuteromethyl iodide (CD_3I) is used, the corresponding macroline trideuteromethiodide (8/25) is formed. This salt decomposes thermally in the mass spectrometer to give macroline and CD_3I as well as CH_3I and d_3-macroline by desalkylation (see Section 8.3). The molecular weights of the compounds 8/18 and d_3-8/18 differ by 3 amu and two overlapping mass spectra are obtained. The results are shown in Table 10. The peaks at m/z 338, 320, and 197 are accompanied by others with an extra 3 amu at m/z 341, 323, and 200 with about the same intensities, but the other four signals do not show this duplication. From this it may be deduced that the two fragments with m/z 320 and 197 contain the basic nitrogen with the methyl group, but not other ions. Moreover two different types of fragmentation must occur. One comprises the ions at m/z 338, 320, and 197, and the other at m/z 338 (or 320), 251, 208, 181, and 170. The latter two ions cannot be formed from the m/z 197 ion, since this would have to lose NH_2 in order to produce m/z 181, which is in contradiction to the mass difference found with d_3-8/18. The same applies to the ion of mass 170.

Making use of various clues, we should next seek to determine the mass spectrometric reaction sequence that leads to the individual ions and thereby derive their structures.

3 A metastable peak indicates that the ion of mass 208 is formed from m/z 251, and high resolution shows that C_2H_3O is split off. That this corresponds to the C-acetyl group of the α,β-unsaturated ketone is revealed by a comparison of the spectra of macrolinol (8/19) and 8/18, in which the keto group has been reduced to an alcohol. In the latter spectrum one finds the heavier ion at m/z 253, but the lighter one is unchanged at m/z 208. Thus $CHOHCH_3$ has been eliminated instead of $COCH_3$.

4 It is extremely unlikely that the ion at m/z 208 is the precursor of the ion at m/z 181, since $CH{=}CH_2$ would have to be split off as a radical. Furthermore, in the spectrum of 20,21-dihydromacroline (8/23) in which the vinyl double bond has been hydrogenated, the macroline ion at m/z 208 is shifted to m/z 210, but the m/z 181 ion is unchanged. Consequently the 181 ion must be formed from m/z 251, 338 (or 320).

5 The ion at m/z 251 with the elemental composition $C_{17}H_{17}NO$ is a radical ion.[12] Thus it must have been formed by a "neutral process".[13]

[12]Molecules that contain C,H,N, and O and are uneven in molecular weight contain an uneven number of nitrogen atoms. In the latter case, when fragmentation occurs, those fragments with uneven mass are radical ions (⁺̇) and those with an even mass are cations (+).
[13]Neutral processes in mass spectrometry are those reactions in which a neutral fragment is formed, such as CO, H_2O, C_2H_4, and $COCH_2$. Reactions in which fragments of this kind are formed include retro-Diels-Alder reactions, McLafferty rearrangements, onium reactions, and S_Ni-type reactions.

• *m/z* 197   The ion containing the smallest residue that has not yet been clarified is *m/z* 197. It thus seems the most suitable candidate to be investigated first. The part that is still unknown comprises 39 amu corresponding to C_3H_3. Summarizing the results so far obtained, we note that the ion at 197 contains the following structural elements:

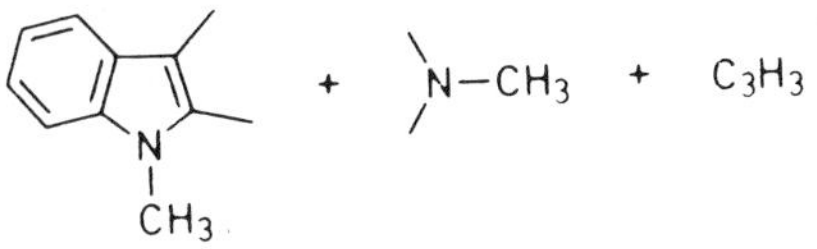

The direct attachment of the indole nucleus to the second nitrogen atom can definitely be excluded on the basis of the UV spectrum of macroline (8/18). It must thus be assumed that in the ion 197 there is a C-C and not a C-N junction with the indole nucleus. Two structures are then possible for this ion, a and b:

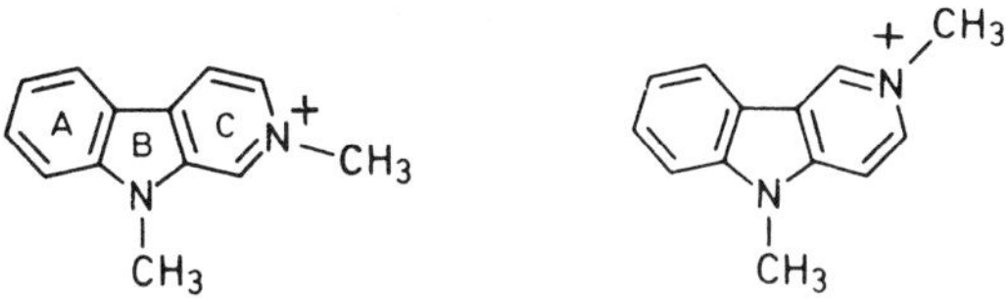

a   *b*

(*m/z* 197)

It is not possible to draw a distinction between these structures on the basis of the experimental results so far obtained, but one can make use of an argument that gives a clue in favor of a. If structure a for the *m/z* 197 ion is used to derive the structure of macroline, a β-carboline system results, which is a tryptamine derivative. In the case of b, a γ-carboline system results. So far only derivatives of β-carboline have been isolated from *Alstonia* spp. Furthermore, β-carbolines preponderate in nature to the extent of more than 99.9% over γ-carbolines. It is thus extremely probable that macroline (8/18) is also a β-carboline derivative, and a is to be preferred over b.

The ion a is derived from the molecular ion of macroline and therefore from macroline itself. However, this residue cannot be incorporated as such into the structure of macroline since it has a quaternary nitrogen and an aromatic ring C. From the UV spectrum it is much more likely that the macroline residue corresponding to a is a tetrahydro-β-carboline, and thus the partial structure II for macroline can be extended to III.

m/z 170   Now that three additional carbon atoms have been fixed in the structure of macroline, an attempt can be made to derive a further ion structure. Compounds with a six-membered ring that contains one double bond can undergo a retro-Diels-Alder reaction in the mass spectrometer. If

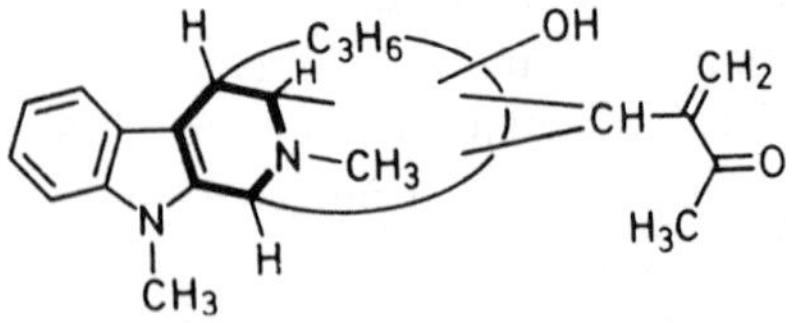

Partial structure III for macroline taking into consideration ion m/z 197.

this occurs for ring C of macroline, the ion so formed would have the partial structure c, which differs from the ion of mass 170 in a minor way only. If $R^1 + R^2 = H + CH_2$, the correct molecular weight and elemental composition results. It follows that two structural alternatives d and e can be written for the ion 170.

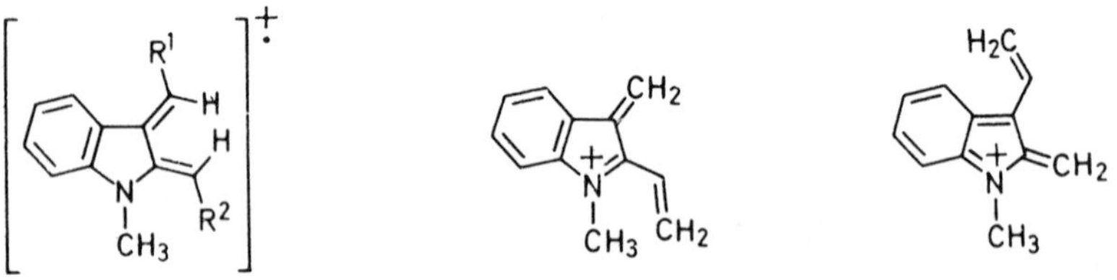

These two structures cannot be distinguished *a priori* on the basis of the chemical and spectroscopic data derived from macroline. However, structure e can be excluded on the basis of analogy with other indole alkaloids (see the previous discussion concerning the ion m/z 197). We can now extend the partial structure of III for macroline and obtain structure IV.

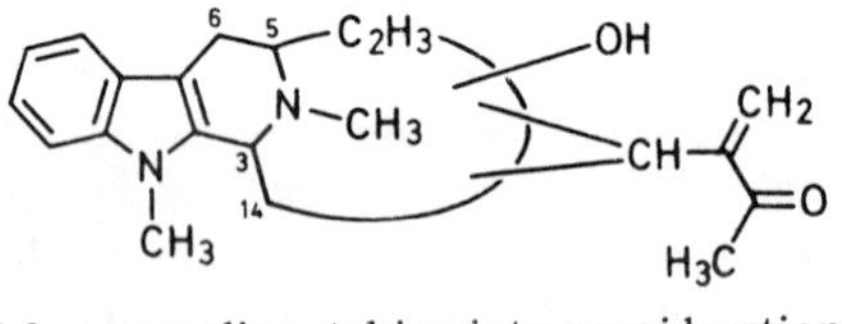

Partial structure IV for macroline, taking into consideration ions 197 and 170.

- m/z **251** As we have seen, the keto group and the indole chromophore are present in ion 251, but the basic NCH_3 and the alcohol group are lacking. We must therefore seek some method by which the NCH_3 and the OH groups might be removed during the formation of ion 251 from macroline. The retro-Diels-Alder reaction suggests itself as one possibility for the elimination of the basic NCH_3 group, but while the bond between C-5 and the rest of the molecule will thereby be broken, the bond with C-14 will be retained.

If we are going to use the retro-Diels-Alder reaction to explain the formation of ion 251, we must connect C-14 to the residue containing the keto group. The resulting partial structure f contains all the C, N, and O atoms required

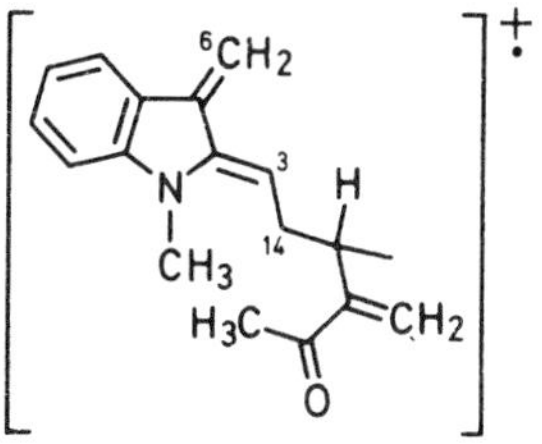

f

for the elemental composition of ion 251, but there is one hydrogen too many and one bond is not saturated. It seems logical to assume that in the fragmentation leading to the m/z 251 ion, this hydrogen atom (at C-14) is lost with the formation of a double bond, which then becomes part of a multiply unsaturated system. The extended partial structure for macroline (8/18) can thus be represented as V.

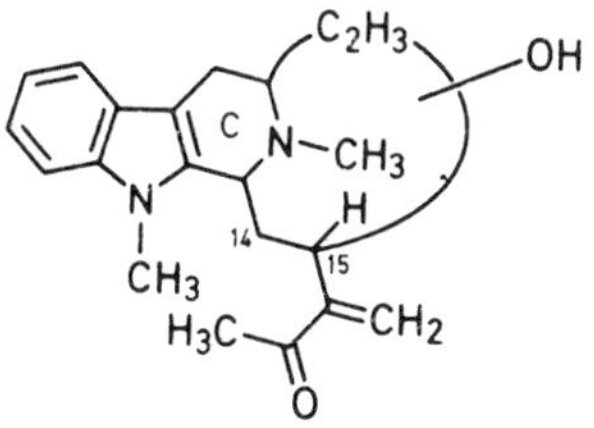

Partial structure V for macroline, taking into consideration ions 251, 197, and 170.

A further insight into the structure is gained from a reconstruction of the method of formation of ion 251. The assumption that a retro-Diels-Alder reaction takes place in ring C of the macroline molecule is evidently a useful one, but the removal of a hydrogen atom from C-14 and the cleavage of the bond between C-15 and the unknown residue of the molecule cannot be explained by a reaction of this sort. The primary retro-Diels-Alder fragmentation must be followed by a neutral reaction[14] in which the hydrogen atom on C-14 is eliminated. One possibility for this reaction is a McLafferty rearrangement, with the $CH_3N=C$-5 bond formed in the retro-Diels-Alder reaction acting as hydrogen acceptor.

---

[14]See footnote 12: m/z 251 is a radical ion containing an odd number of nitrogen atoms, so its mass must also be odd.

Since McLafferty rearrangements originate from a structure with a suitable arrangement of six atoms, the fragmentation pattern shown in Scheme 19 is required. A seven-membered arrangement, which might be written by

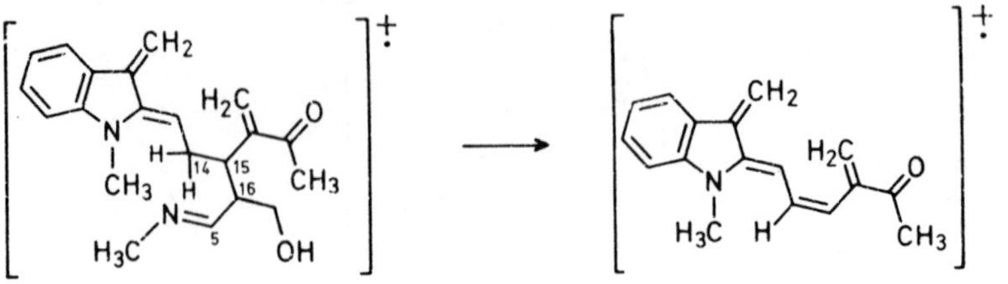

g
(m/z 251)

Scheme 19   Formation of the fragment m/z 251 from macroline (8/18).

incorporating the second unknown carbon atom, is extremely unlikely. Moreover, the hydroxy group cannot be located at C-16, otherwise an additional (C)—CH_3 would be detected in the NMR spectrum. From these considerations the (planar) structure 8/18 is arrived at for macroline.

The complete fragmentation mechanism for macroline (8/18) is shown in Scheme 20. The formation of ions 208, 181, and 320 requires some clarification.

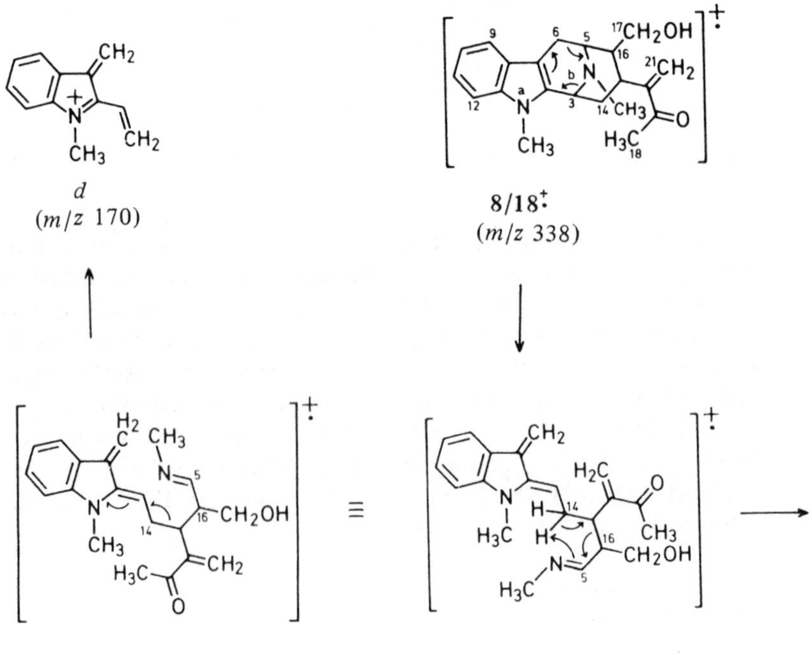

d
(m/z 170)

8/18$^{+\cdot}$
(m/z 338)

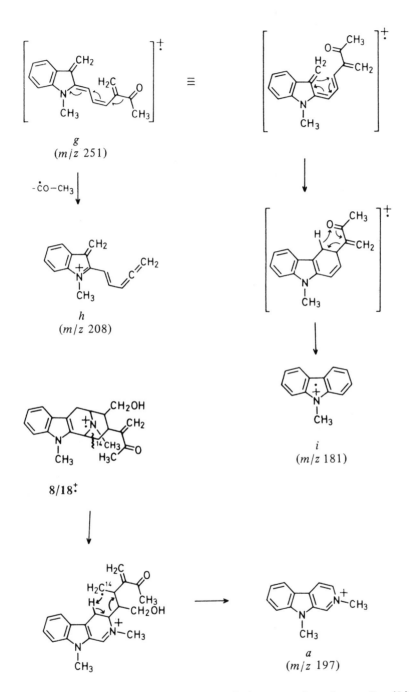

Scheme 20 Complete mass-spectrometric fragmentation of macroline (8/18).

The ion m/z 208 arises from m/z 251 ($m*$) by cleavage adjacent to the 20, 21 double bond. A position adjacent to a C,C double bond is usually unfavorable for a cleavage, but in the above-mentioned case it leads to a stabilized acetyl radical and to the conjugated ion m/z 208. The formation of ion m/z 181 can be explained only by the pathway shown (Scheme 20). However, the structure and formation of this ion admittedly lend no support to the macroline structure, since the ion has a carbazole and not a β-carboline skeleton. Starting from macroline, the following types of fragmentation take place in sequence: retro-Diels-Alder reaction, McLafferty rearrangement, Diels-Alder reaction, and finally another McLafferty rearrangement.

The loss of water from the molecular ion has so far not been completely explained. Presumably this reaction takes place from the hemiacetal form (8/18a):

8/18a

The line of reasoning we have followed depends essentially on spectroscopic arguments and leads to the structure 8/18 for macroline, which is supported by the spectroscopic properties of synthetic model compounds [H 21], and of similar alkaloids occurring in the same plant [K 10]. However, this structure for macroline is not completely proved. Further chemical studies would be necessary for this purpose and for clarifying the stereochemistry of the compound. If we are, however, prepared to assume that the structures of the two halves of the molecule are now known, we can venture to tackle the structure of villamine (8/16) next, and finally that of villalstonine (8/1).

8.6 STRUCTURAL ELUCIDATION OF VILLAMINE (8/16) AND VILLALSTONINE (8/1)

As we have seen in some detail, on pyrolysis of villamine (8/16) the bases pleiocarpamine (8/14) and macroline (8/18) are formed (see Scheme 21). The molecular weights of these two bases added together give the molecular weight of villamine, and thus nothing but these two compounds is formed in the pyrolysis reaction. We can exclude small fragments such as H_2O and CO_2, which are formed in other pyrolysis reactions, and we only need to take into consideration compounds 8/14 and 8/18.

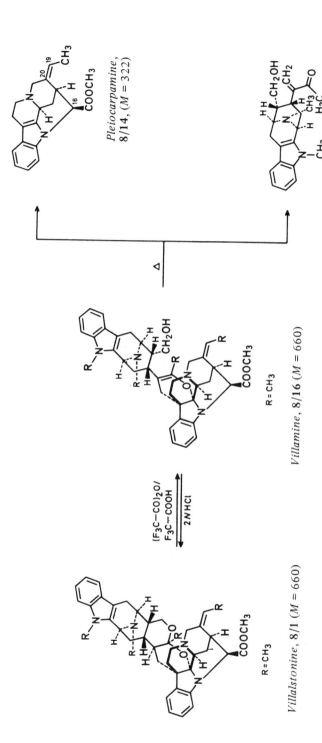

Scheme 21 Transformation products of villalstonine, 8/1.

139

The positions in the "monomeric" bases at which they are joined together to form the "dimer" are most easily deduced from a comparison of their functional groups with those of villamine (8/16). Any groups that are additional or are missing, or which are changed in some way, give a clue as to the position of junction. The functional groups of the three compounds are set out in Table 11,

Table 11   Comparison of Functional Groups in the Alkaloids Pleiocarpamine (8/14) and Macroline (8/18) with those of Villamine (8/16)

Group	Pleiocarpamine (8/14) + Macroline (8/18)	Villamine (8/16)
COOCH$_3$	1 (8/14)	1
CH$_2$OH	1 (8/18)	1
C=O	1 (8/18)	None
C=COR	None	1
Basic N	2 (8/14 + 8/18)	2
Aromatic H	4 + 4	8
Chromophore	Indole + indole	Indole + indoline
Ethylidene group	1 (8/14)	1
Vinyl group	1 (8/18)	None
Methyl group	1 NCH$_3$ (8/18)	1 NCH$_3$
	1 COCH$_3$ (8/14)	None
	1 COOCH$_3$ (8/14)	1 COOCH$_3$
	1 C=CCH$_3$ (8/14)	2 C=CCH$_3$

from which it can be seen that the α,β-unsaturated ketone function of macroline (8/18) must be involved in the linkage, as well as the double bond in the indole nucleus of pleiocarpamine (8/14) since these groups are missing in villamine. The latter, however, has an enol ether double bond (IR: 1682 cm^{-1}) and an additional (C)$-$CH$_3$ (singlet, 1.31 ppm) as well as an indoline chromophore.[15] Furthermore it must be remembered that 8/14 and 8/18 are formed by an especially facile thermal reaction. Consequently there are two possible modes of formation of the mixture of macroline and pleiocarpamine:

[15]If one subtracts the UV spectrum of macroline from that of villamine (Figure 3) one obtains an indoline spectrum, the same as that of 2,7-dihydropleiocarpamine (8/15). Conversely, the spectrum of villamine is obtained by combination of the UV spectra of macroline and 2,7-dihydropleiocarpamine (Figure 3). The ^{1}H-NMR spectrum shows that the indole chromophore of pleiocarpamine is involved and not that of macroline. In the spectra of both 8/18 and 8/16 the indolic N-methyl group appears at the same chemical shift, which would not be expected in the case of a 2,7-dihydromacroline chromophore.

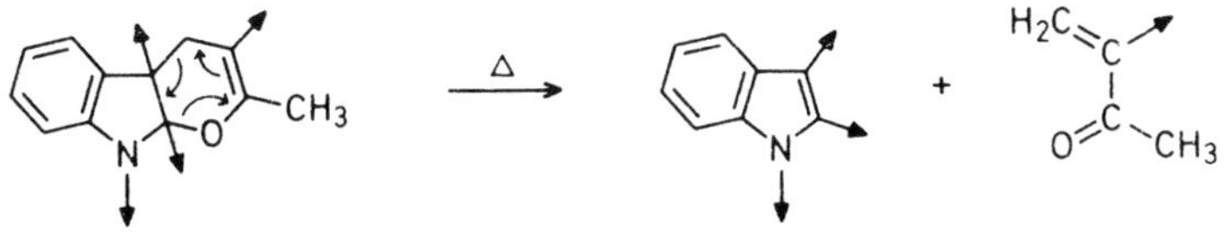

Villamine
First possibility
(partial structure)

Pleiocarpamine

Macroline

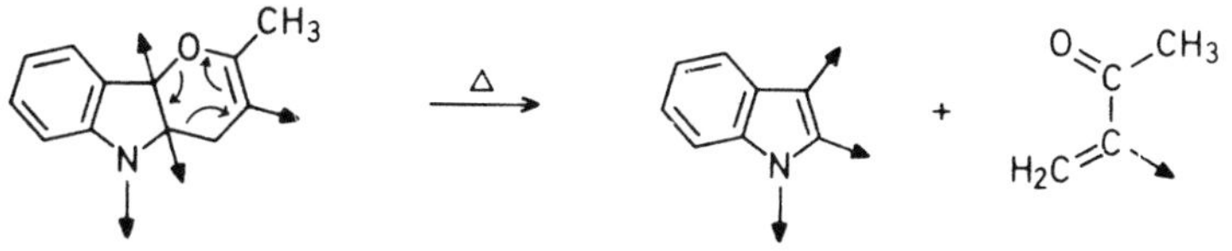

Villamine
Second possibility
(partial structure)

Pleiocarpamine

Macroline

It is not possible to decide between these alternatives without further evidence, but if one considers the probable biogenetic formation of villamine from macroline and pleiocarpamine, the first possibility would be preferred, since the most electronegative position in the indole nucleus, apart from the nitrogen, is the β-position; the α-position is less favored:

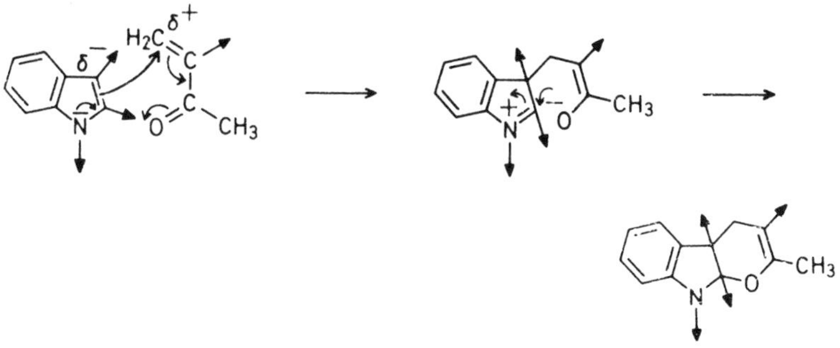

Villamine (partial structure)

This leads to structure **8/16** for villamine, in which the α-*cis*-arrangement of the macroline portion on carbon atoms 2 and 7 of pleiocarpamine is preferred in analogy to the position of the hydrogen atom in 2,7-dihydropleiocarpamine. A β-arrangement is not possible on steric grounds. The final step involving the derivation of the structure of villalstonine (**8/1**) itself results, as before, from a consideration of the functional groups. While villamine can form an *O*-acetyl derivative and shows an enol ether band in its IR spectrum, both groups (CH_2OH, $C=C-O-R$) are lacking in villalstonine, which on the other hand

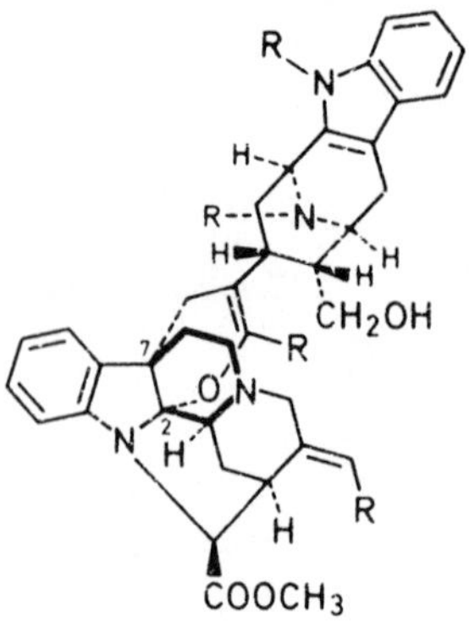

R = CH₃

8/16, *Villamine*

shows the presence of an additional $(C)CH_3$ singlet in its ^{1}H-NMR spectrum (at 1.23 ppm; in villamine this appears at 1.31 ppm). From these data the structure of villalstonine is represented by **8/1** [H 21]. The isomeric base villoine (**8/17**)

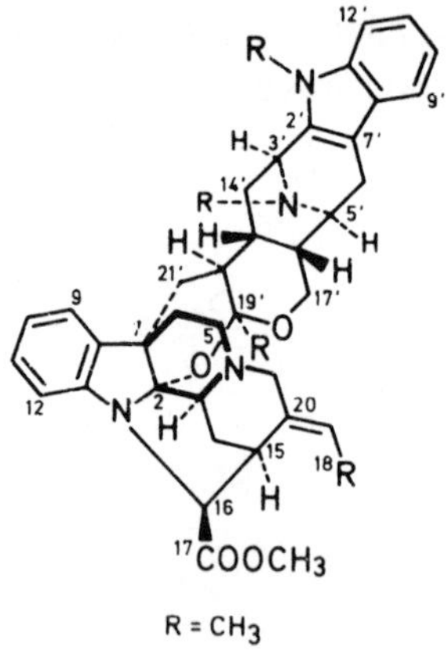

R = CH₃

8/1, *Villalstonine*

presumably has a structure analogous to **8/16**, but with the C-2—O—C-19' bond opened. The structure thus deduced for villalstonine has been confirmed by X-ray crystallographic analysis [N 3], and at the same time the stereochemistry at centers 3', 5', 15', 16', 19', and 20' of the macroline portion was also determined.

Lack of space has prevented us from undertaking a detailed interpretation of all the spectra, but it should be emphasized that a comprehensive study of this kind is essential in order to confirm the structure. For the most part we have been interested only in the more prominent absorption bands that were relevant to our discussion on the structural elucidation of macroline and villalstonine.

The discussion on the structural determination of villalstonine illustrates the power of modern spectroscopic methods. The successful application of mass spectrometry to this problem depends essentially on the fact that macroline gives an especially clear fragmentation, commencing in ring C of the molecule. The derivatives **8/19**, **8/20**, and **8/23** show a completely analogous behavior and likewise proved to be useful in the structural analysis.

An analogous approach can be used for the structural elucidation of other alkaloids. The most fruitful spectroscopic method to apply in each case depends entirely on the class of substance to which the alkaloid belongs. If the type of nucleus in the alkaloid cannot be deduced from its spectra alone as in the case of villalstonine, one or more degradative reactions must be sought which will permit an insight into the structure of the alkaloid.

Important Degradation Reactions of Alkaloids

Chemical degradations continue to play an important role in the structural eluci-dation of alkaloids. Their object is the formation of smaller molecules whose structures are easier to recognize, or products with fewer rings or other func-tional groups. Typical examples of a general kind include ozonolysis, periodate cleavage, and permanganate oxidation. Reactions of these types have been successfully applied in the structural elucidation of many other compounds which do not contain nitrogen and which will not be dealt with here. We shall confine our attention to reactions that are characteristic of alkaloids and require the presence of an amino nitrogen, such as the Hofmann, Emde, and von Braun degradations. In the following discussion, general aspects of these degradation reactions and their special features will be discussed.

9.1 HOFMANN DEGRADATION

The Hofmann reaction plays a key role in the degradation of alkaloids. It is rela-tively easy to carry out, and usually the structural changes in the resulting prod-uct as compared to the original alkaloid are easily recognizable by spectroscopic analysis. Although the Hofmann degradation is used less frequently compared to the time when the structural elucidation of alkaloids was carried out entirely by classical methods, its utility is nevertheless undiminished in cases where modern spectroscopic methods do not provide an unequivocal answer (e.g., spermidine and spermine alkaloids). The reaction involves the conversion of a tetraalkylammonium compound into a tertiary amine and an olefin by the action of base. For the formation of the quaternary ammonium compound, a primary, secondary, or tertiary amine is treated with an excess of methyl iodide.

The ammonium iodide derivative (methiodide) is then reacted with either a strong or a weak base according to the type of compound. The reaction conditions vary considerably, and the following examples are typical:

1 Exchange of I^- for OH^- on an ion exchange column, with or without heating [M 37]

2 Treatment of the ammonium salt (usually chloride or hydroxide) with NaOH, $NaOCH_3$, or $KOC(CH_3)_3$, and so on [P 15]

3 Heating of the ammonium iodide with Ag_2O in aqueous or alcoholic solution [S 44]

4 Heating the dry methofluoride (see Section 8.3 under Thermal Hofmann degradation) or methohydroxide in high vacuum (with distillation of the product) [D 4], [H 22], [V 1]

Some cases are known in which a Hofmann degradation takes place merely by thin-layer chromatography of the quaternary iodide with a weakly basic solvent for elution, so that it is not possible to give reaction conditions that will be generally applicable. In order to protect the product of the reaction, it is best to start with the mildest reaction conditions that will be effective.

A necessary condition for a Hofmann degradation is the presence of a hydrogen atom in a β-position to the quaternary nitrogen. Usually the elimination follows a E2 mechanism in which the removal of the H^+ and the N group is simultaneous, but one often finds a tendency toward other reaction types. Thus if the $C-N^+$ bond starts to break first, the α-carbon atom acquires a carbonium ion character, and the elimination mechanism approaches the E1 type. On the other hand the elimination of the β-hydrogen atom can take place somewhat earlier than the splitting of the $C-N^+$ bond, resulting in a mechanism that approaches the E1cB type. From these alternatives it is evident that the position of the double bond that is formed will be determined according to either the Saytzeff rule or the Hofmann rule. The latter predicts that the least substituted olefin will be formed. The reaction will follow the Saytzeff rule if the mechanism has a more E1 character, while the preferential formation of an α-carbonium ion will be in accord with the Hofmann rule. This will result in the removal of the most acidic hydrogen attached to a β-carbon atom, and thus the hydrocarbon with the smallest number of substituents will result.

Steric and electronic factors alike are concerned in the formation of the product on Hofmann degradation. The steric arrangement of the atoms taking part in the reaction (β-hydrogen and ammonium group) cannot be defined in simple terms. Examples have shown that the antiperiplanar as well as the synperiplanar conformation can lead to the formation of the product. Furthermore, by incorporation of a β-deuterium instead of a β-hydrogen atom it has been possible to detect in special cases an E_i mechanism, indicating a cis arrangement of the atoms in question; for example,

$$[R_2CD-CH_2-\overset{+}{N}(CH_3)_3]\,OH^- \longrightarrow R_2CD-CH_2-\overset{\overset{\displaystyle {}^-CH_2}{|}}{\underset{+}{N}}(CH_3)_2 \longrightarrow$$

$$H_2CD-N(CH_3)_2$$

Evidently the mechanism is closely dependent on both the alkaloid structure and the reaction conditions. Mixtures of various products can be detected as a result of varying the latter. The general equation for the reaction is as follows:

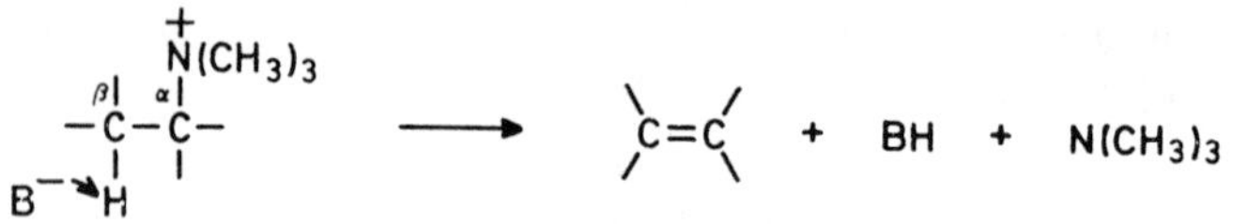

$$-\overset{\overset{\displaystyle \overset{+}{N}(CH_3)_3}{|}}{\underset{\beta|}{C}}-\overset{\overset{\displaystyle \alpha|}{|}}{\underset{|}{C}}- \longrightarrow \quad \overset{}{\underset{}{C}}=\overset{}{\underset{}{C}} \quad + \quad BH \quad + \quad N(CH_3)_3$$
$$B^{-\,\curvearrowright}H$$

Various possibilities arise according to the number of bonds between the nitrogen atom and the rest of the molecule. If the nitrogen in the original alkloid is joined by one bond only, a neutral olefin and an amine are produced. The behavior of thalicthuberine [9/1, from *Thalictrum thunbergii*, (RAN)] may be taken as an example of this type of reaction, which results in the conversion of 9/1 into its methine[1] 9/2 (see [F 5]).

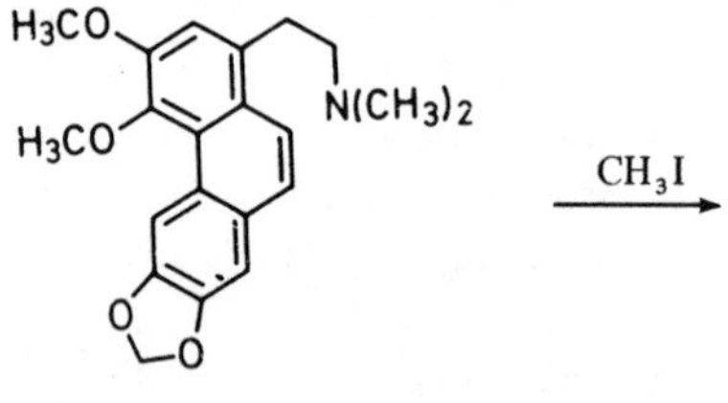

9/1, *Thalicthuberine*

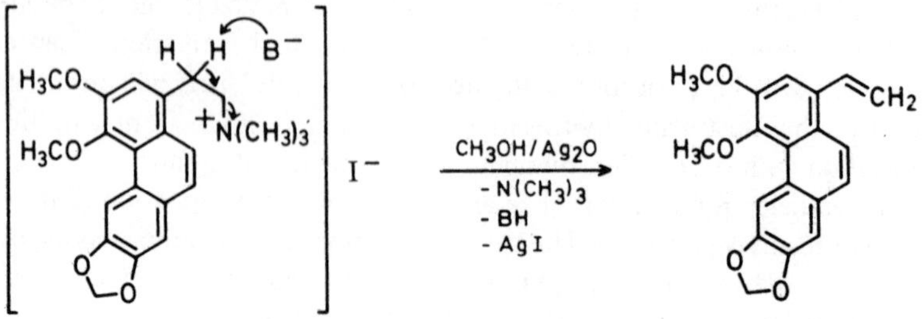

9/2

[1] The olefin formed by Hofmann degradation is referred to as a Hofmann base or methine.

If the amine is cyclic, the Hofmann degradation leads to an opening of the ring, which happens in the case of the quaternary alkaloid magnocurarine [9/3, from various species of the genus *Magnolia* (MAG)]. Treatment of 9/3 with CH₃I/KOH leads at first to the quaternary base 9/4 in which the phenolic hydroxy groups are methylated. Under the influence of KOH this is converted into the methine 9/5[2], a stilbene derivative [M 38].

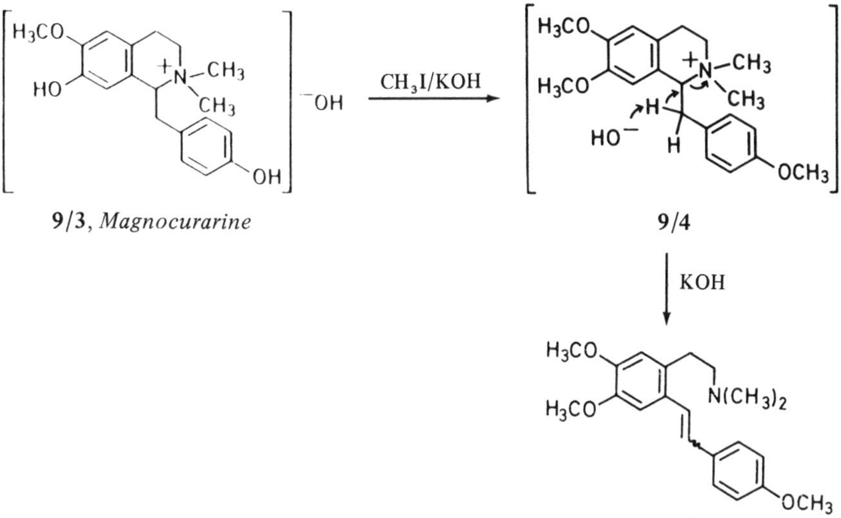

9/3, *Magnocurarine*

9/4

| KOH

9/5[2]

If on the other hand the amino nitrogen forms part of two rings simultaneously, only one of them is opened by the Hofmann reaction. Thus the product on Hofmann degradation of tylophorine (9/6, see Section 5.1.11) is tylophorine methine (9/7), in which the nitrogen forms part of a central ring system [G 13], [G 14].

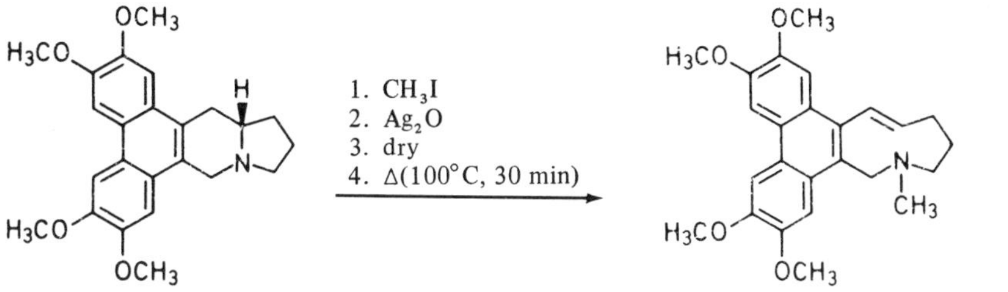

9/6, *Tylophorine*

1. CH₃I
2. Ag₂O
3. dry
4. Δ(100°C, 30 min)

9/7

[2]The main product is the trans isomer; the cis isomer is also formed in smaller amounts.

In cases where nitrogen is still present in the methine that is formed initially, a nitrogen-free methine can finally be obtained by repeated methylation and Hofmann degradation (exhaustive methylation). The number of degradations required reveals the number of bonds between the nitrogen and the rest of the carbon skeleton. In order to avoid undesirable side reactions, the C,C double bonds formed in the degradations are often reduced catalytically between the various degradative steps.

If we assume that none of the factors dealt with suosequently in this section influences the position of the olefinic double bond that is formed, then of the various possibilities, the olefin with the smallest number of alkyl groups will be formed (Hofmann rule), that is, the olefin with the least number of substituents.

If the β-hydrogen is in a benzyl group, it is removed preferably as shown in examples 9/1, 9/3, and 9/6. Double bonds and carbonyl groups also exert an activating influence. In the first case a conjugated double bond is formed [e.g., 3-phenyl-*N*-methylgranatenine (9/8)] and in the second an α,β-unsaturated carbonyl compound (e.g., the phenyl ketone 9/9). In general the most acidic hydrogen is removed, and in the case of alkyl-substituted compounds this will occur where the smallest number of substituents is located.

Apart from these examples, other products of Hofmann degradation can be formed as described below in cases where special structural features are present.

9.1.1 Hofmann Degradation with Formation of Epoxides (Oxirane Formation)

If a hydroxy group occurs in a β-position to a quaternary nitrogen atom, the latter can be eliminated under the basic condition of the Hofmann degradation with the simultaneous formation of an epoxide. Thus the alkaloid conhydrine [9/10, from *Conium maculatum* (UMB)] forms the epoxide 9/11 after methyla-

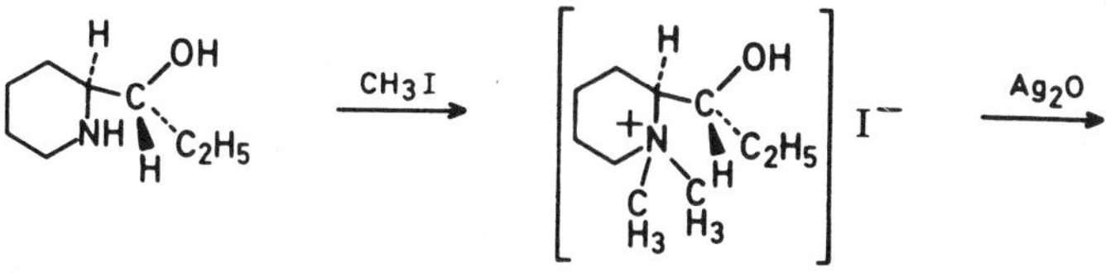

9/10, (+)-*Conhydrine*

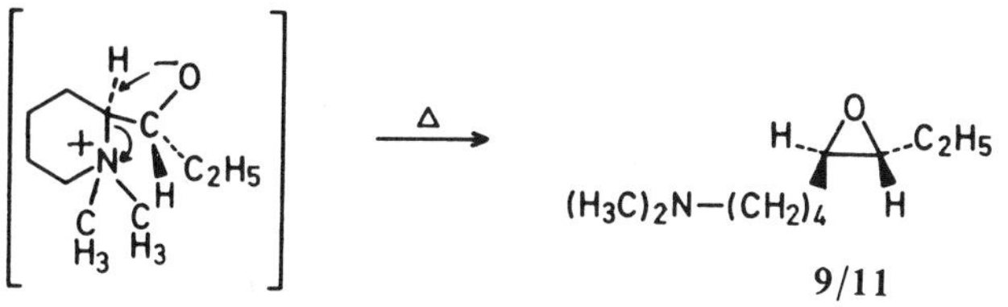

9/11

tion and Hofmann degradation [S 44]. The quaternary base does not undergo Hofmann degradation directly, but first forms the alkoxide as shown, which then determines the course of the reaction. Other alkaloids which behave in this way include (−)-ephedrine and (+)-pseudoephedrine [9/12 and 9/13, respectively, from various *Ephedra* spp. (EPH), among others.] [W 13].

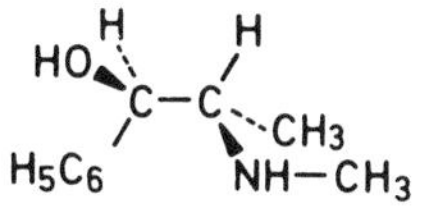

9/12, (−)-*Ephedrine*[(1R,2S)-(−)-2-methylamino-1-phenylpropan-1-ol]

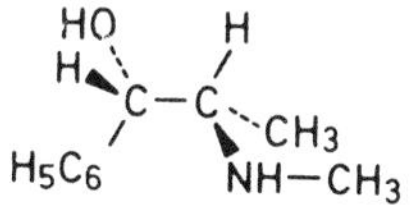

9/13, *Pseudoephedrine*[(1S,2S)-(+)-2-methylamino-1-phenylpropan-1-ol]

9.1.2 Hofmann Degradation with Ether Formation

A β-hydroxy group is not the only requisite for the formation of a cyclic ether. Hydroxy groups further removed from the nitrogen can also be directly involved. α-Erythroidine [9/14, from various *Erythrina* spp. (LEG)] is degraded in several steps to the base 9/15, which on Hofmann degradation forms the tetrahydro-furan derivative 9/16 instead of the expected base 9/17 [G 15]. A nucleophilic substitution reaction takes place here as well instead of an elimination. Similar degradation reactions leading to cyclic ethers have been exploited in the structural elucidation of chelidonine [9/18, from *Chelidonium majus* (PAP) among others] [B 30] and cryptopine [9/19, from various species of the genera *Corydalis*, *Dicentra*, and *Fumaria* (FUM) as well as *Eschscholtzia*, *Macleaya*, and *Papaver* (PAP)] [P 15]. An interesting feature of the degradation of 9/19 is the

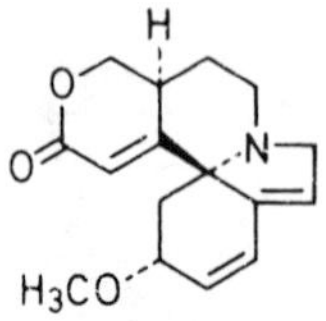

9/14, α-*Erythroidine* 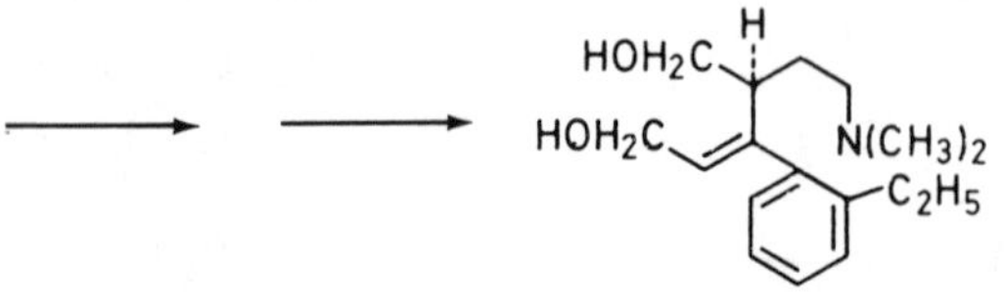   **9/15**

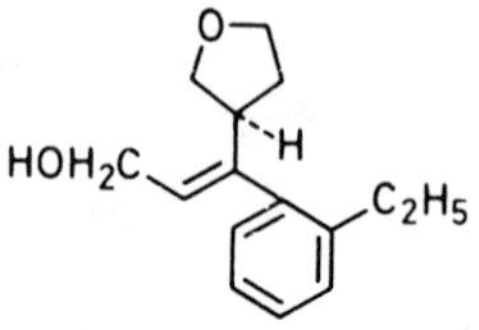

9/16 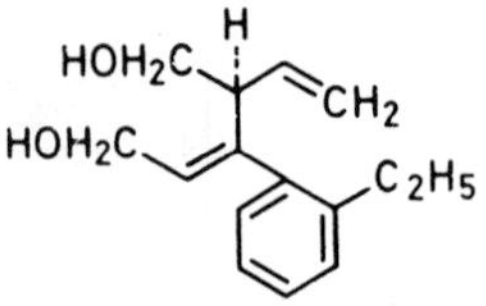   **9/17**

appearance of two methine bases. In addition to the normal base **9/20**, the cyclization product **9/21**, formed from the enolic base, is obtained [P 15].

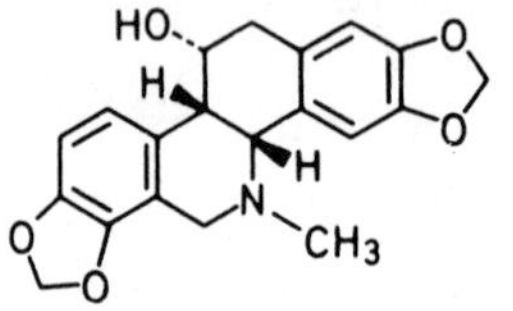

9/18, (+)-*Chelidonine*

1. CH₃I
2. Hofmann degradation

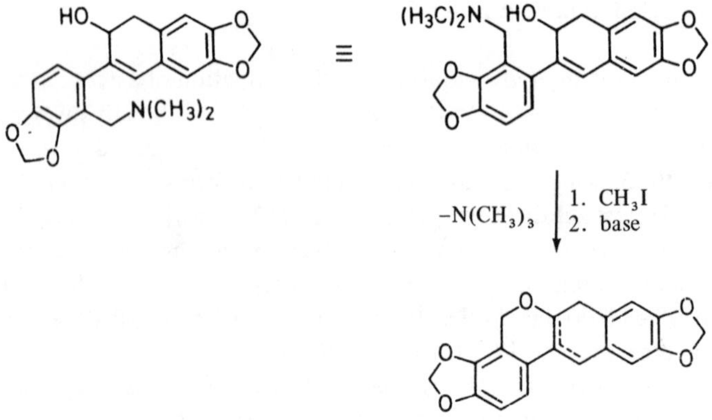

1. CH₃I
2. base

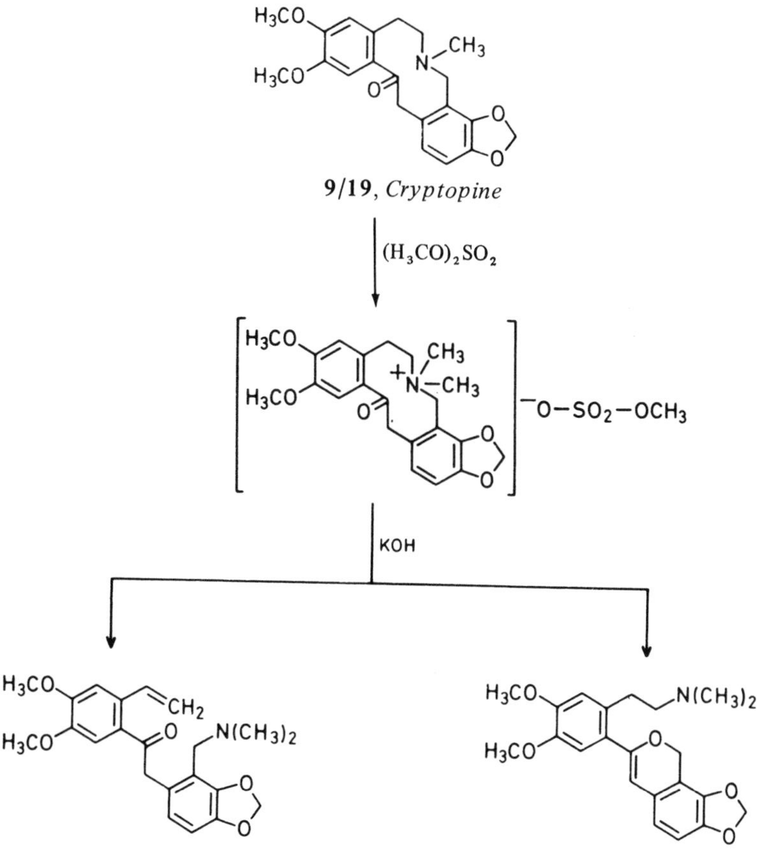

9/19, *Cryptopine*

$(H_3CO)_2SO_2$

$^-O-SO_2-OCH_3$

9/20

9/21

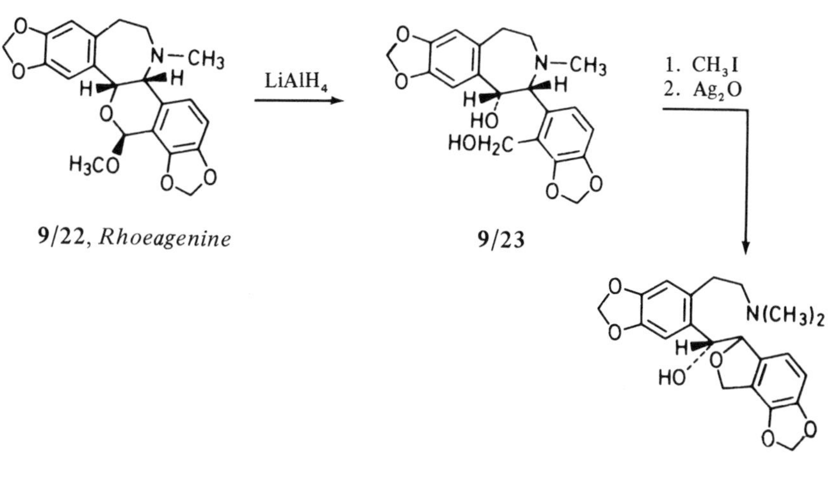

9/22, *Rhoeagenine*

LiAlH$_4$

9/23

1. CH$_3$I
2. Ag$_2$O

151

In this connection attention should also be drawn to the Hofmann degradation of rhoeagenindiol (9/23), a reduction product of rhoeagenine [9/22, from *Papaver* sp. (PAP); see Section 5.1.7.5] ([S 46] and references cited therein).

9.1.3 Further Reactions under Hofmann Degradation Conditions

Apart from cyclic ethers, other products can be formed in a Hofmann degradation which likewise do not involve an elimination. The spermidine alkaloid *N*-methylcodonocarpine [9/24, from *Codonocarpus* sp. (GYR)] has one basic and two amido nitrogens. If the quaternary compound *O*-ethyl-*N'*,*N'*-dimethyl-tetrahydrocodonocarpine hydroxide (9/25) is heated, the *N*-acylpyrrolidine derivative 9/26 is formed by attack of the N″ atom on the α-carbon to the N′, instead of the expected product 9/28. The base 9/26 can be converted by means of a second methylation and degradation [D 4] into the "normal" product 9/27 through loss of trimethylamine.

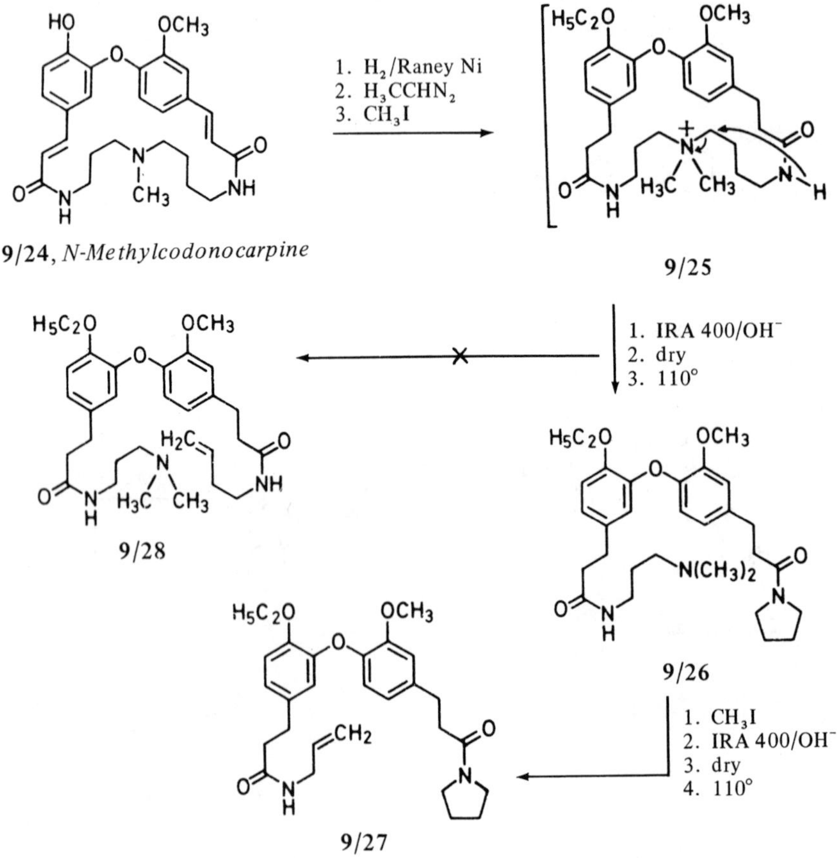

9/24, *N-Methylcodonocarpine*

1. H_2/Raney Ni
2. H_3CCHN$_2$
3. CH$_3$I

9/25

1. IRA 400/OH⁻
2. dry
3. 110°

9/28

9/26

1. CH$_3$I
2. IRA 400/OH⁻
3. dry
4. 110°

9/27

[3]The two phenyl ether groups are distinguished by treatment with diazoethane.

The Hofmann degradation of the indole alkaloid 19,20-dehydroervatamine [9/29, from *Ervatamia* sp. (APO)] is of special interest. In a preliminary reaction the methoxycarbonyl and keto groups are reduced with LiAlH$_4$, then the secondary alcohol group is reoxidized with CrO$_3$/pyridine to give the ketone 9/30. The first Hofmann reaction involves a vinylogous elimination in which a δ-hydrogen atom is removed to produce 9/31. A hydrogen atom in the β-position is lacking for the second Hofmann degradation as well, but there is a hydroxy group present instead. The proton is removed from the latter by the strong base (see Sections 9.1.1 and 9.1.2), and after loss of formaldehyde and trimethylamine (retro-Mannich reaction, fragmentation reaction) the product 9/33 is formed, which rearranges under conditions of basic catalysis into the α,β,γ,δ-unsaturated ketone 9/34 [K 11].

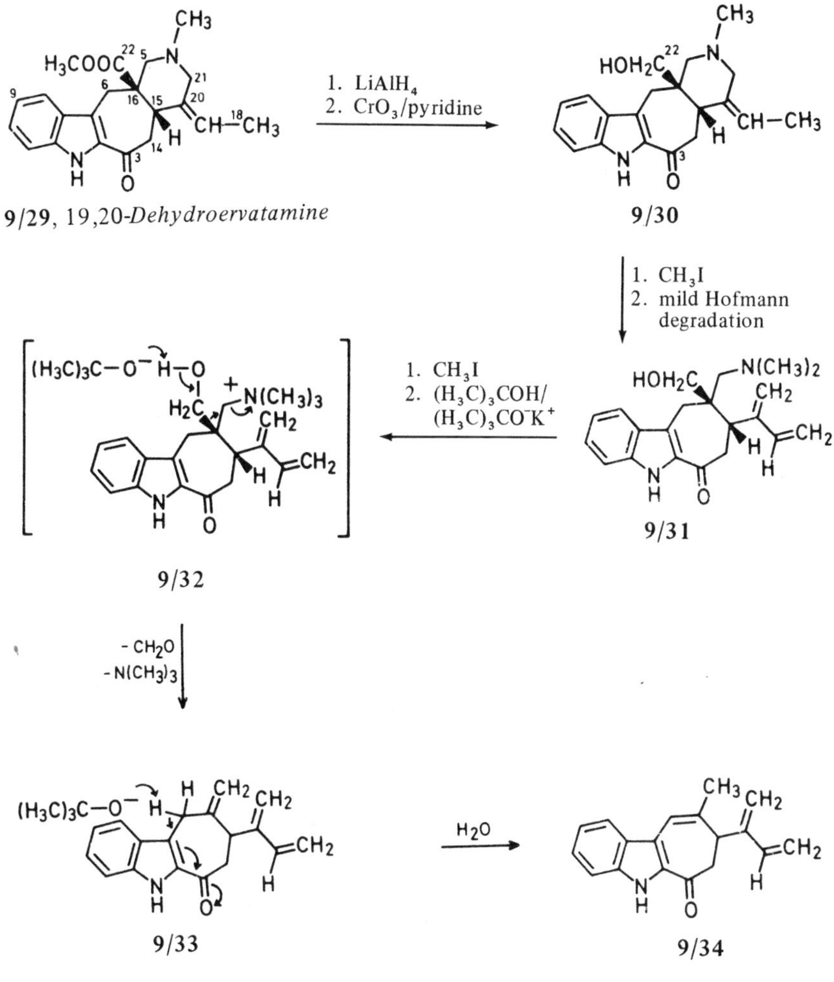

9/29, 19,20-*Dehydroervatamine*

9/30

9/32

9/31

9/33   9/34

It may be mentioned in this connection that under the very strongly basic conditions a number of other reactions similar to those involved in the formation of 9/34 from 9/33 can take place, which are not connected with the actual Hofmann degradation.

The presence of certain structural features can also lead to an ethylogous Hofmann degradation. Codeine [9/35, from *Papaver* spp. (PAP); see Sections 5.1.7.16, and 6] is converted by Hofmann degradation into the base 9/36, which after treatment with methyl iodide and distillation of the quaternary hydroxide yields the so-called methylmorphenol 9/38. The first stage of the reaction involves the elimination of water to form 9/37, followed by a Hofmann degradation in which the phenanthrene derivative 9/38 is formed together with ethylene and trimethylamine [M 39]. The driving force in this reaction is no doubt the formation of the resonance-stabilized aromatic system.

1. $(H_3CO_2)SO_2$
2. NaOH/Hofmann degradation

9/35, (−)-*Codeine*   9/36

Methylation and
Hofmann degradation

X^-

$-N(CH_3)_3$
$-H_2C=CH_2$

9/37

9/38

Methylmorphenol

As we have seen, the Hofmann degradation is dependent on the presence of at least one hydrogen atom in a β-position, either directly or vinylogously, to a quaternary nitrogen. Various substitution reactions compete with this elimination process. They may be intramolecular, or they may involve attack of the base at an α-carbon (e.g., attack of OH⁻ with formation of an alcohol) and can lead to the removal of the quaternary nitrogen atom. Apart from substitutions, base catalyzed reactions that can take place under the conditions of the Hofmann degradation such as epimerizations, retro-Mannich reactions, and retro-aldol reactions must also be kept in mind. The multiplicity of these reactions might appear at this juncture to be a source of confusion rather than of clarification, but it must be remembered that these variations are almost exclusively dependent on the presence of other functional groups. When one considers a range of different reactions occurring under various reaction conditions, it becomes clear that any one type of functional group is seldom capable of one reaction only—for the most part each group can undergo several different reactions—so that the Hofmann degradation is not exceptional in this respect.

9.2 EMDE DEGRADATION

A further reaction depending on the presence of a quaternary nitrogen is the Emde degradation, which can be effected under a variety of reductive conditions. The reaction is often carried out in aqueous or alcoholic solution with sodium amalgam, but Emde degradations have also been recorded involving the use of catalytic hydrogenation or sodium in liquid ammonia.

One of the C-N$^+$ bonds is cleaved in the course of the reaction, and a tertiary amine is formed together with a saturated alkyl residue:

$$[R-CH_2-\overset{+}{N}(CH_3)_3]\, X^- \xrightarrow[-HX]{[2H]} R-CH_3 + N(CH_3)_3$$

Of the four different bonds attached to the quaternary nitrogen atom, preferential cleavage takes place of those that are activated through attachment to benzylic or allylic groups. Thus the metho salt of tylophorine (9/6) is converted into isodihydrohomotylophorine (9/39) by treatment with sodium amalgam [G 16]. Various reaction products or mixtures of them are obtained according

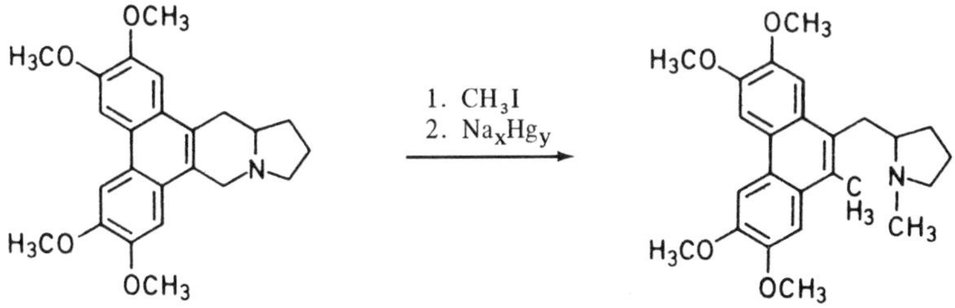

9/6, *Tylophorine*   9/39

to the reaction conditions used. This kind of behavior is shown by a derivative of pseudobrucine [9/40, from *Strychnos* sp. (STR)]. By treatment with CH$_3$OH/H$^+$, the alkaloid is transformed into *O*-methylpseudobrucine through a *trans*annular reaction:

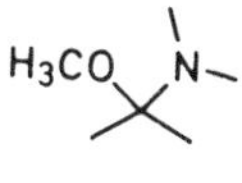

A subsequent reaction with methyl iodide forms the quaternary salt, which gives an *N*-methylpseudobrucine enol ether with sodium methoxide by a Hofmann reaction:

OCH3

N—CH3

The enol ether is quaternized with methyl iodide to give **9/41**. On reduction with sodium amalgam this yields **9/42**, in which only the allylically activated bond N-4—C-21 has been reductively cleaved. Catalytic hydrogenation of **9/41** also takes place with fission of this bond, but the isolated 19,20 double bond is subsequently hydrogenated as well to produce **9/43** [B 21].

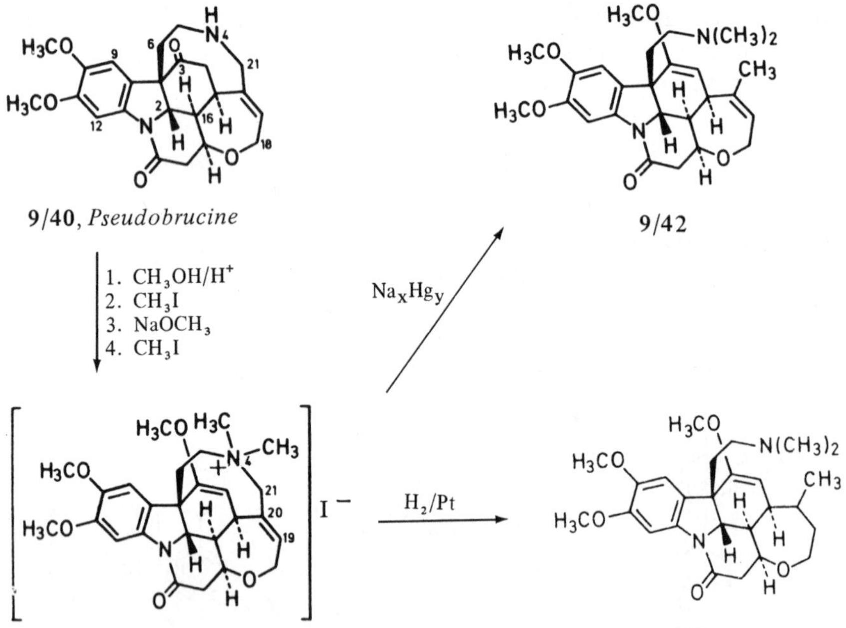

9/40, *Pseudobrucine*

1. CH_3OH/H^+
2. CH_3I
3. $NaOCH_3$
4. CH_3I

Na_xHg_y

9/42

9/41

I^- H_2/Pt ——→

9/43

The Emde degradation of C-fluorocurine (**9/44**; see Section 5.1.2.3.4) follows a similar course to the formation of **9/43** from **9/41**. The carbonyl group in the yellow indoxyl chromophore of **9/44** is reduced with sodium borohydride to the corresponding colorless alcohol **9/45**, which on catalytic hydrogenation yields the so-called ϵ_1-hexahydrofluorocurine (**9/46**), accompanied by cleavage of the N-4—C-21 bond and hydrogenation of the 19,20 double bond [B 31].

On the other hand the Emde degradation takes quite a different course in the case of C-mavacurine [**9/47**, from calabash curare and *Strychnos* sp. (STR)], an alkaloid closely related to C-fluorocurine (**9/44**). Instead of the expected cleavage of the allylic N-4—C-21 bond, the C-3—C-4 bond, which is also allylic, is ruptured instead. On acidification of **9/48** the quaternary compound **9/49** is obtained, which can undergo a reversible Hofmann degradation [B 31]. This interesting type of reaction is generated by the ten-membered ring present in **9/48**.

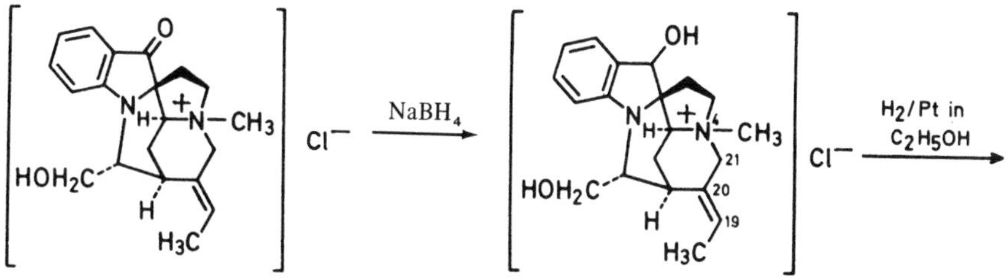

9/44, *C-Fluorocurine*   9/45, *Hydrofluorocurine*

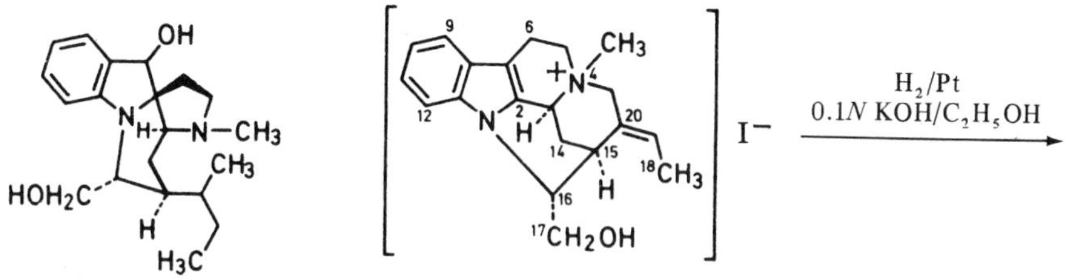

9/46, ϵ_1-*Hexahydrofluorocurine*   9/47, *C-Mavacurine*

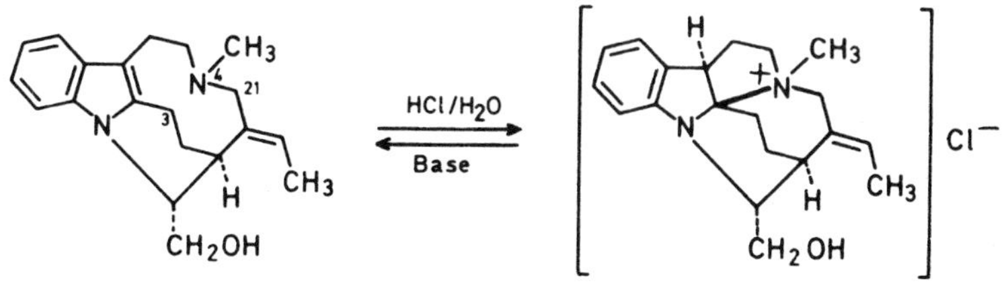

9/48, ϵ_2-*Dihydromavacurine*   **9/49**

Substances that have neither benzylic nor allylic groups attached to the quaternary nitrogen atom can nevertheless be degraded by the Emde reaction. Conessine (**9/49a**; see Section 5.6) can be converted in several stages (exhaustive methylation, Hofmann degradation, catalytic hydrogenation) into tetrahydro-conessinemethine dimethiodide (**9/50**), which is reduced with Na/liquid NH_3 to a mixture of 3-β-dimethylaminoallopregnane (**9/51**) and allopregnane (**9/52**).

The cleavage of the N—C-18 bond takes place in the formation of both degradation products, but the reaction involving the 3-amino group differs in the two cases. For the formation of 9/51 the N—CH₃ bond is broken, and for 9/52 the N—C-3 bond [H 23].

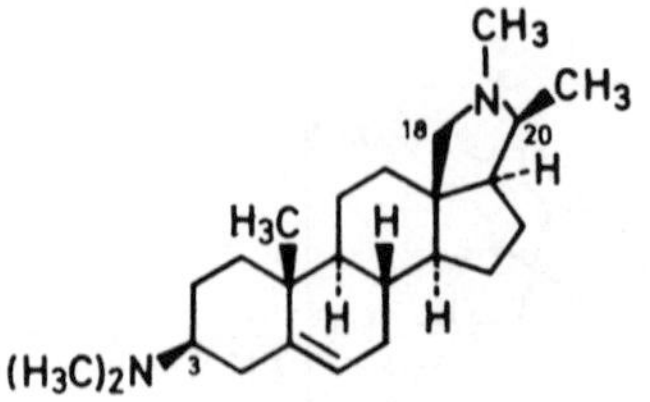

9/49a, *Conessine*

1. CH₃I
2. KOH
3. H₂/cat.
4. CH₃I

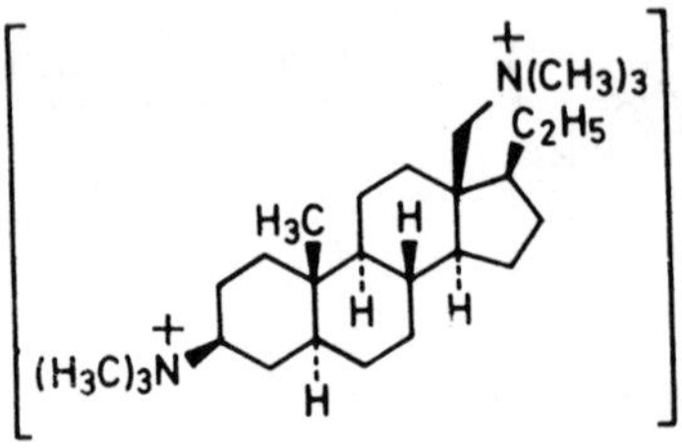

9/51

9/50

Na/liquid NH₃

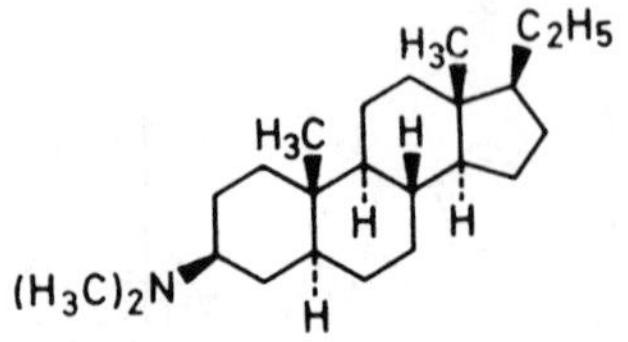

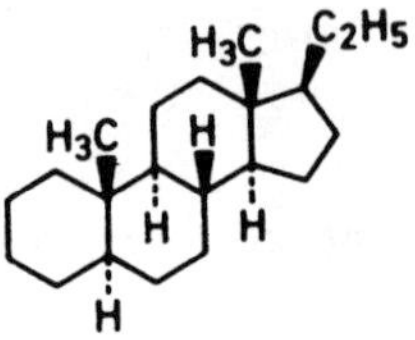

9/52, *Allopregnane*

A final example that may be mentioned is the Emde degradation of the O-methyldimethine 9/53, which is formed by a two-stage Hofmann degradation from (−)-cryptaustoline [9/54, from *Cryptocarya* sp. (LAU)]. In this case the reaction involves the rupture of a phenylamine bond to form 9/55 [E 1].

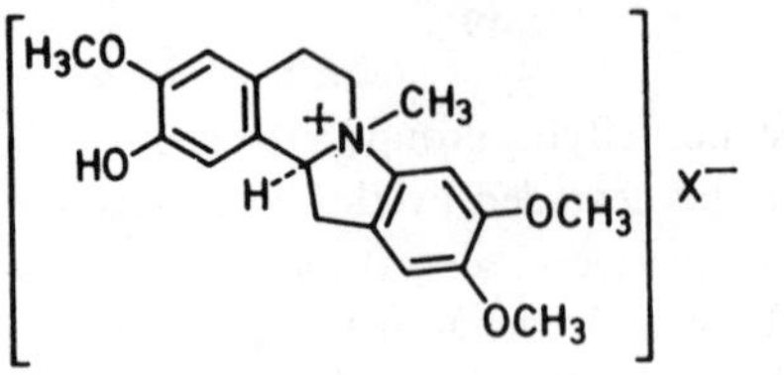

9/54, *Cryptaustoline*

1. CH₃I/KOH
2. 2 Hofmann degradations

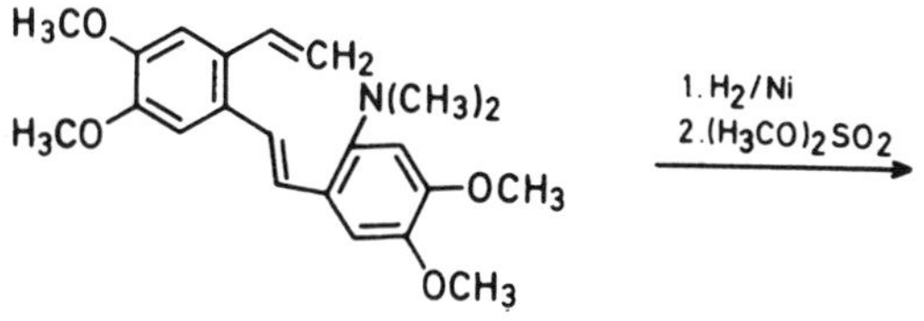

9/53

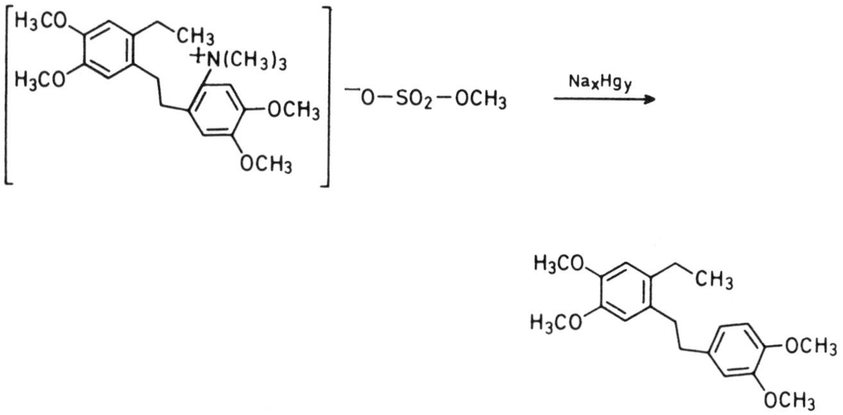

9/55

These few examples of the Emde degradation will suffice to demonstrate the the diversity of the reaction, which is now seldom used but still has a certain importance in the catalytic hydrogenation of naturally occurring quaternary nitrogen compounds.

9.3 THE VON BRAUN REACTION (CYANOGEN BROMIDE DEGRADATION)

When a tertiary amine is reacted with cyanogen bromide (BrCN), an alkyl bromide and a disubstituted cyanamide are formed. Thus on treatment of the ester amine 9/56 (desoxy-N-methylcarpamic acid ester) with BrCN, three different cleavage reactions may in principle be expected. Only one main reaction in fact takes place, the formation of 9/57 and methyl bromide [G 16]. The removal of the nitrile group from the cyanamide residue is usually carried out by acid-catalyzed hydrolysis to the corresponding carbamic acid, which decarboxylates spontaneously to form the secondary amine. The cyanamide can also be converted into the secondary amine with lithium aluminum hydride [P 16].

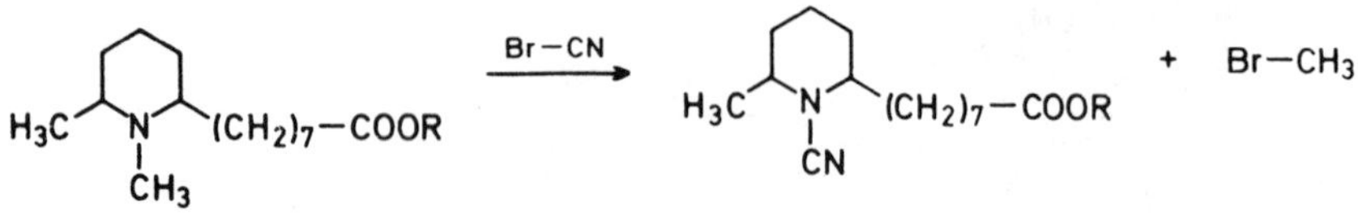

9/56   9/57

The successful use of the von Braun reaction is illustrated in the following examples. The pyrrolizidine derivative 9/58 reacts with cyanogen bromide to form 9/59:

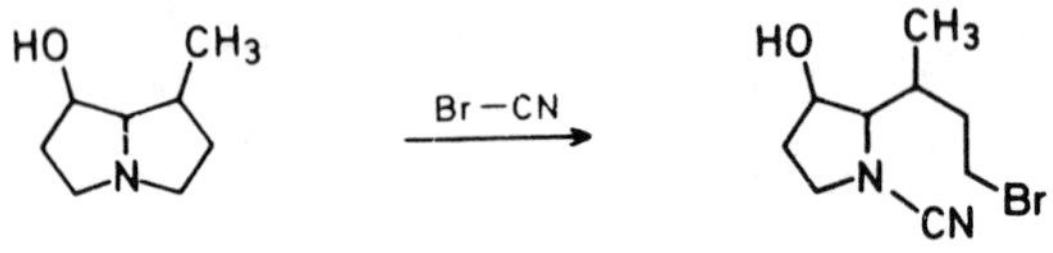

9/58   9/59

The von Braun degradation was applied successfully in the structural elucidation of lycopodine [9/60, from various *Lycopodium* spp. (LYC); see Section 5.1.11.3.4). The reaction afforded two products which are formed by opening either the C-9—N-14 bond or the C-1—N-14 bond [M 40]. The same method of degradation has also been used in the structural determination of other *Lycopodium* alkaloids.

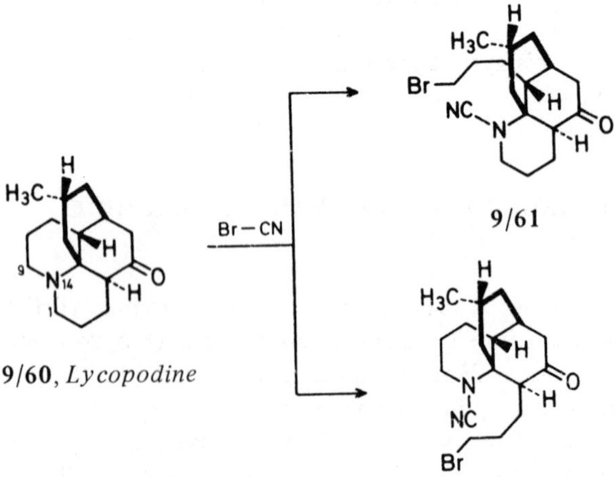

9/61

9/60, *Lycopodine*

9/62

The substance dihydroerysotrine (9/63) is formed by the hydrogenation of erysotrine, a trimethyl ether that so far has not been found to occur in nature.

It is formed by methylation of erysodine (9/64), erysovine, or erysopine [all from various *Erythrina* spp. (LEG)]. The cyanogen bromide degradation of 9/63 gives a product that after treatment with lithium aluminum hydride is converted to the diphenyl derivative 9/65 [P 16].

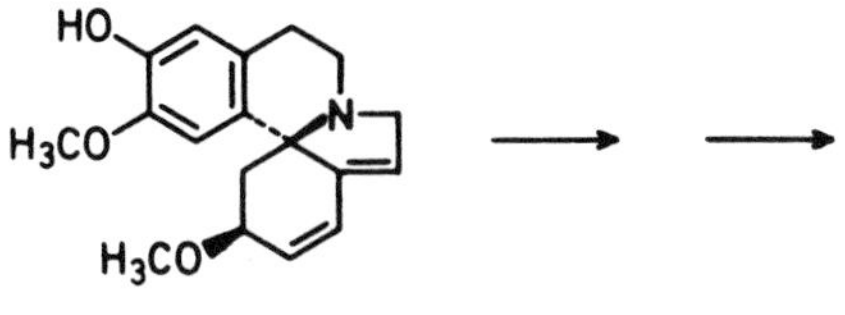

9/64, (+)-*Erysodine*

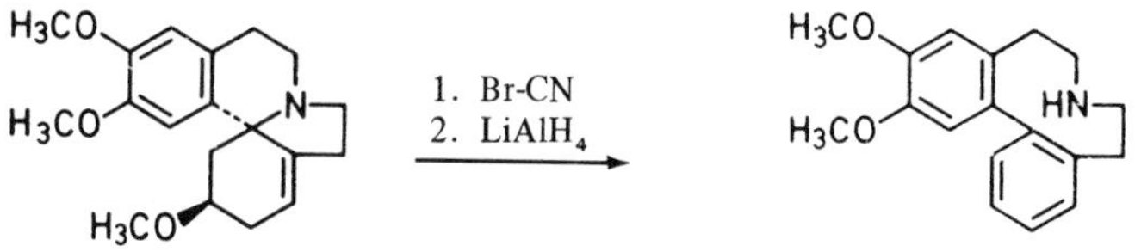

9/63   9/65

Before closing this chapter we may well ask why three different degradation reactions involving the amino group are in use, and how they come to maintain their existence. Apart from the fact that the von Braun degradation starts off from a tertiary amine only, it often happens that one or the other of these reactions fails in a particular case or gives a number of products that are difficult to separate. The use of a different degradation reaction may then lead to success.

Dimeric Alkaloids
Bisalkaloids

As we have seen from the classification of alkaloids (Section 5), a vast range of skeletal types is encountered among naturally occurring plant bases. However, this by no means exhausts the complete range of alkaloids. There are also naturally occurring bases that comprise two of these simple units with a direct linkage between them. They are often referred to as "dimeric" in contrast to the "monomeric" alkaloids from which they are built up, although this method of naming them is not always correct. The name bisalkaloid as a general term for the whole range of binary bases is sometimes inappropriate as well, but we shall nevertheless make use of it in the absence of a better designation.

There are at present about 300 representatives of this group of compound, of which about 120 belong to the bisindole and a similar number to the bisiso-quinoline alkaloid types. The remaining bases are distributed throughout all the usual structural types. There is thus a distinct preference for indole and iso-quinoline alkaloids, which is perhaps connected with the fact that an especially large number of plants containing the monomeric representatives have been investigated so far. There are also a few (ca. 10) alkaloids known in which three bases of similar structural type are joined together in the same molecule. Finally, there are two alkaloids known that are formed from the union of four building blocks of the same kind. It is now apparent that natural inspiration is richer than our previous conception of it, and no doubt there are actually "polymeric" examples. However, problems are encountered due to the relatively high molecular weights, together with the accumulation of polar functional groups, which makes it difficult to detect and isolate such substances contained in plant extracts. In the mass spectrometric investigation of these large molecules, thermal decomposition reactions take place which seriously interfere with the determination of the molecular weight. This class of compound is thus somewhat speculative, and further studies will be required to show what kind of properties these bases have.

Apart from a few exceptions that will be dealt with later, the bisalkaloids are composed of two bases of the same type; the bisindoles [C 3], [G 8], the bisisoquinolines, and the bisquinolizidines are examples. The preferred combination of skeletal types depends on the content of "monomeric" alkaloids in the particular plant. *Catharanthus roseus* (Apocynaceae), probably the most intensively studied of all plants, contains indole alkaloids exclusively, and the bisalkaloids isolated from it are bisindole alkaloids. Similar considerations apply to a range of other alkaloid-bearing plants. It may be concluded that biosynthesis of the bisalkaloids starts off with the synthesis of the "monomeric" representatives, followed by their "dimerization," that is, the union of two units to form a bisalkloid. Often this second stage can be carried out in the laboratory, which permits certain of these substances to be synthesized in good yield. On the other hand by reversal of the method of synthesis, the bisalkaloid may be cleaved into the two halves that go to make it up.

However, the biosynthesis of bisalkaloids does not necessarily start from two complete monomeric units. Other methods of synthesis are conceivable, and so far to our knowledge no direct evidence to distinguish between possibilities has been put forward.

What reactions are involved in the formation of these "dimers"? The following section will deal with the principles involved in these syntheses and will be illustrated by various examples.

10.1 MANNICH REACTION

A large number of alkaloids appear to be formed from monomeric bases by a Mannich reaction, as is evidently the case for the bisindole alkaloid pleiomutine [10/1, from *Pleiocarpa mutica* (APO)]. Of the two halves that make up the alkaloid, one has an eburnamine skeleton and the other a pleiocarpine skeleton (see Scheme 22). Pleiomutine can be cleaved with dilute hydrochloric acid into eburnamenine (10/3) and pleiocarpinine (10/4), both of which occur in *P. mutica*. The synthesis of 10/1 likewise takes place in hydrochloric acid solution from pleiocarpinine and eburnamine (10/2). The eburnameninium ion 10/5 is formed first, in which C-14 is electrophilic. The aromatic centers 15 and 17 of 10/4 are para and ortho, respectively, to the indolic nitrogen and thus bear a partial negative charge. Reaction occurs exclusively at C-15 for steric reasons, and also because of its greater nucleophilicity. The aromatic character of the benzene ring in the pleiocarpinine portion is reestablished by loss of a proton with formation of 10/1. The formation of a large number of other bisindole alkaloids such as voacamine (10/6, from various genera of the Apocynaceae) can be explained by an analogous mechanism. The latter base is accessible synthetically from the *Iboga* alkaloid voacangine (10/7) and from vobasinol (10/8) by

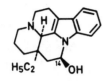

10/2, (–)-*Eburnamine*

$-H_2O$ | H_3O^+

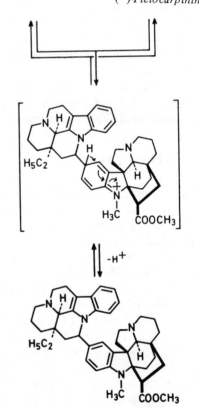

10/3,
Eburnamenine

−H⁺

10/5

10/4,
(–)-*Pleiocarpinine*

−H⁺

10/1, Pleiomutine

Scheme 22 Possible mode of biogenesis of pleiomutine (**10/1**).

164

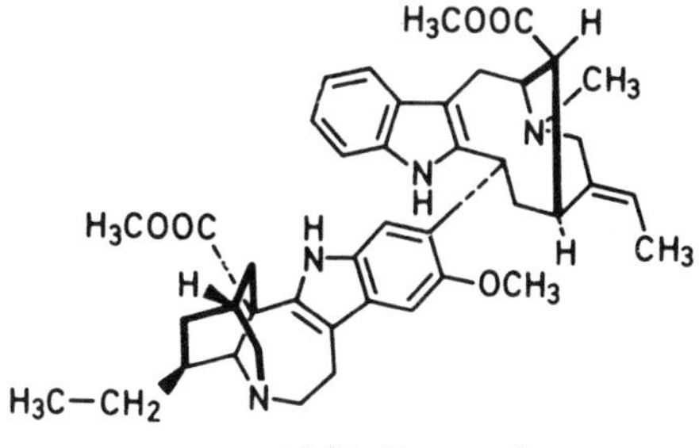

10/6, *Voacamine*

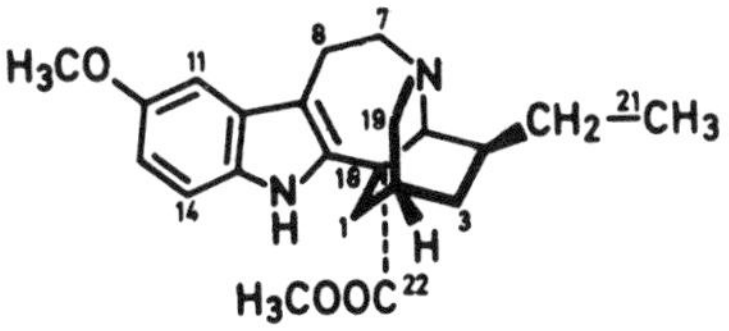

10/7, *Voacangine*

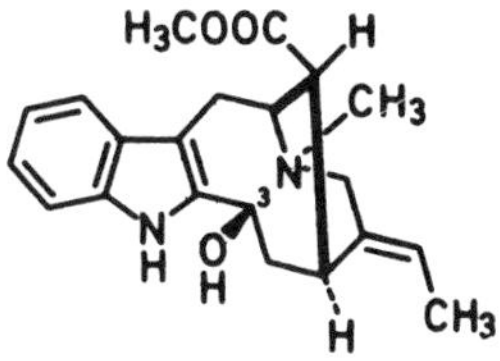

10/8, *Vobasinol*

treatment with dilute hydrochloric acid. During the reaction C-3 in **10/8** becomes electrophilic. The alkaloid sanguidimerine [**10/9**, from *Sanguinaria* sp. (PAP)] is a dimer of the benzophenanthridine alkaloid sanguinarine chloride (**10/10**; see **5/144**) and acetone in 2:1 molar proportion, and is formed by a double Mannich reaction. We need not go into the question as to whether this compound is a natural product or—as is more likely—an artefact formed from **10/10** and acetone. The alkaloid dendrocrepine [**10/11**, from *Dendrobium* sp. (ORC)] presents a similar case.

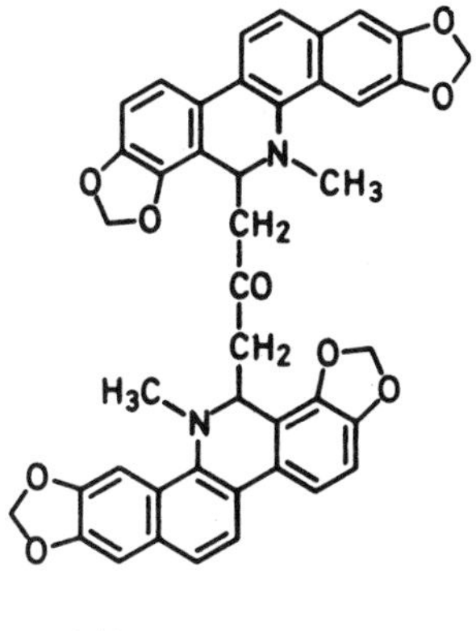

10/9, *Sanguidimerine*

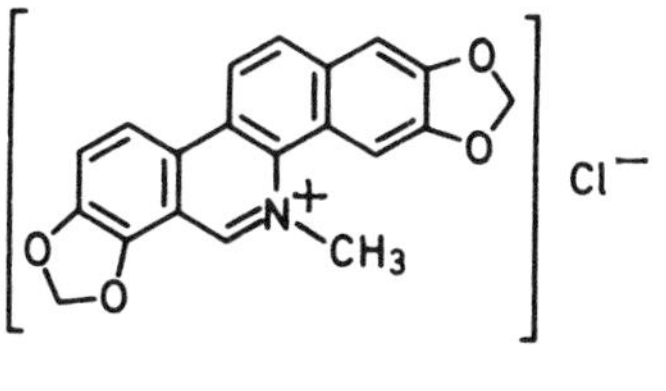

10/10, *Sanguinarine chloride*

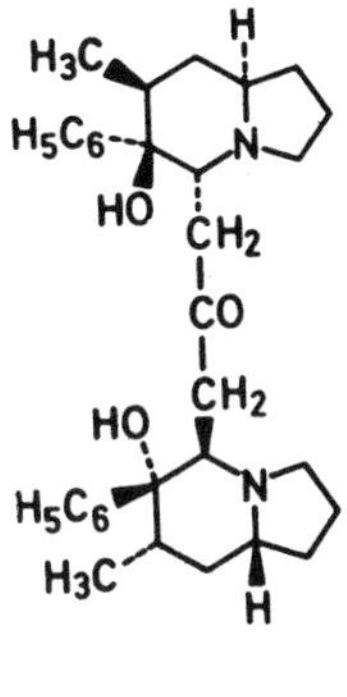

10/11, *Dendrocrepine*

10.2 MICHAEL CONDENSATION

The formation of a number of alkaloids can be ascribed to a Michael condensation, and an example is villalstonine, which was discussed at length in Section 8.

10.3 ALDEHYDE-AMINE REACTIONS

A very large group of dimeric alkaloids is formed by reaction between a primary or secondary amine in one monomeric unit with an aldehyde group in a second unit. The curare alkaloid dihydrotoxiferine [**10/12**, from calabash curare and various South American *Strychnos* spp. (STR)] can be interpreted as being formed by a reaction of this sort [B 10]/[B 11], [B 12], [B 18]. The monomeric precursor is presumably the aldehyde hemidihydrotoxiferine (**10/13**), which is readily transformed into **10/12** by warming in dilute acetic acid solution. The base **10/13** is reformed from **10/12** with dilute mineral acid.

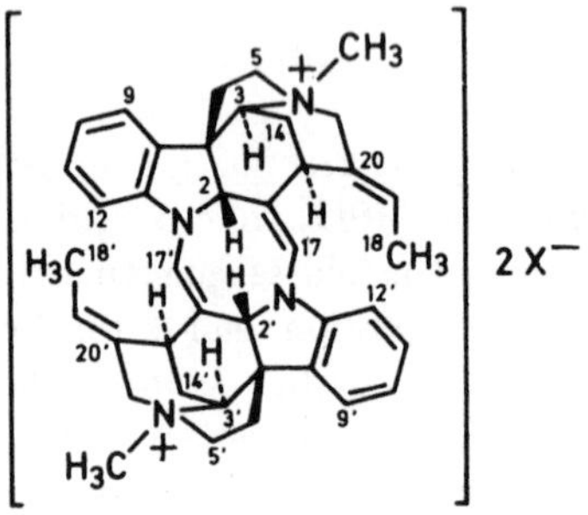

10/12, *Dihydrotoxiferine*

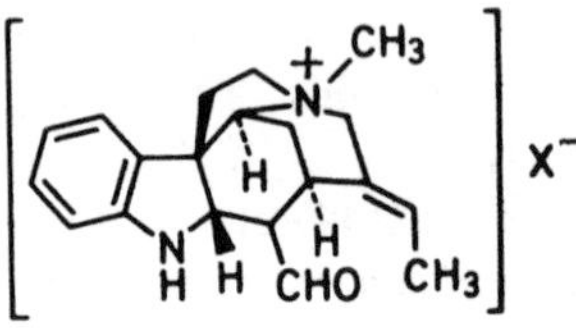

10/13, *Hemidihydrotoxiferine*

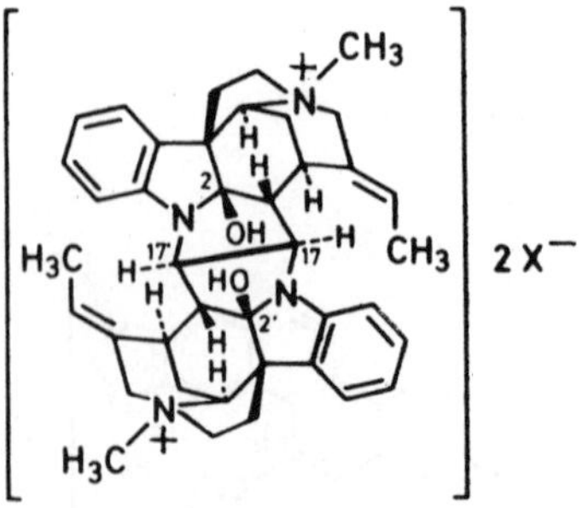

10/14, *Calebassine*

The tendency of these dimeric compounds toward cleavage disappears when the molecule undergoes a change at the position of junction. Thus calebassine (**10/14**), which occurs along with **10/12**, is a naturally occurring oxidation product of the latter, but it is no longer cleaved into two halves on treatment with acid. Calebassine can be obtained in good yield by oxidation of dihydrotoxiferine. It contains two extra hydroxy groups on carbon atoms 2 and 2′ as compared to **10/12**, and moreover the two carbon atoms 17 and 17′ which are involved in the dimerization are joined to one another by a C,C bond which is responsible for the behavior of **10/14** with mineral acid. As a result it is not possible to cleave the molecule without first opening this new C,C bond. Apart from **10/12** and **10/14**, 15 other curare alkaloids of known bisindolic structure have so far been isolated from natural sources. The curare alkaloids are muscle relaxants which interfere with depolarization. The semisynthetic base alloferine (18,18′-dihydroxy-*N,N′*-demethyl-*N,N′*-diallyldihydrotoxiferine) is used in surgery as a muscle relaxant.

While dimeric bases like the curare alkaloids are built up from two complete monomeric bases, alkaloids such as tubulosine [**10/15**, from *Alangium lamarckii* (ALA) and *Pogonopus tubulosus* (RUB)] occupy an intermediate position between dimeric and monomeric bases. Tubulosine is formed from the isoquinoline alkaloid protoemetine [**10/16**, from *Psychotria* sp. (RUB)] and 5-hydroxytryptamine (**10/17**).

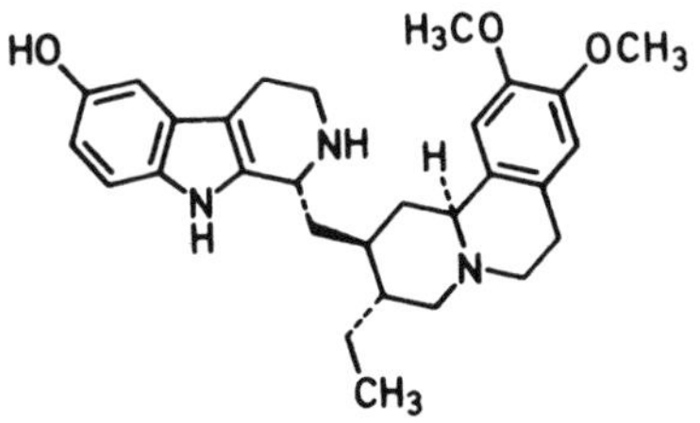

10/15, *Tubulosine*

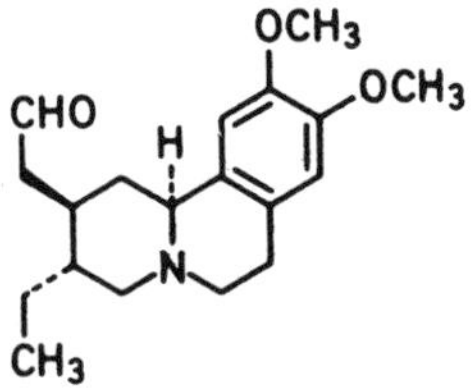

10/16, *Protoemetine*

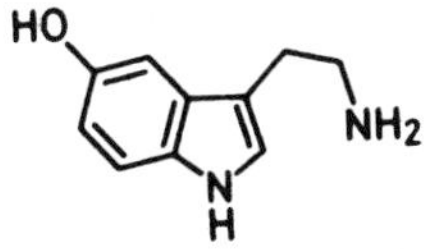

10/17, 5-*Hydroxytryptamine*

The aldehyde group at position 16 of **10/16** and the primary amine group in **10/17** are important for the interaction between the two halves. Apart from the two aromatic building blocks that consist of tryptamine and tyrosine derivatives, tubulosine contains only an extra aliphatic residue. There is a resemblance in structure between the alkaloids of the tubulosine type and the "monomeric" ipecacuanha alkaloids (see Section 5.1.7.6) which, however, have an additional tyramine residue instead of the tryptamine unit. A range of other alkaloids is known which are constructed in a way to similar tubulosine. Another example is roxburghine D [10/18, from *Uncaria* sp. (NAU)].

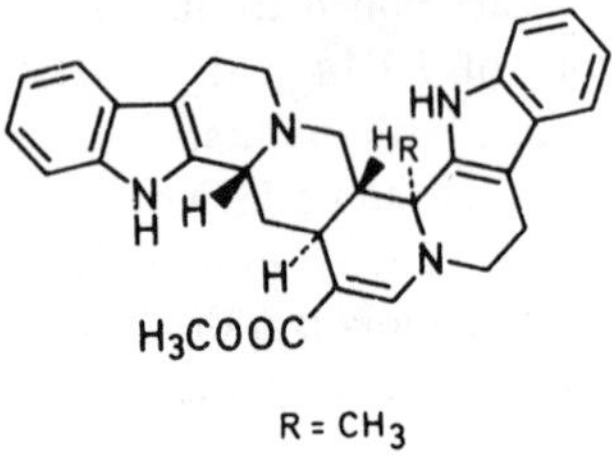

R = CH3

10/18, *Roxburghine D*

10.4 COUPLING BY PHENOL OXIDATION [T 20]

Practically all the 120 or so bisbenzylisoquinoline alkaloids [C 2], [C 17], [K 1], [K 2] owe their formation, at least hypothetically, to oxidative phenolic coupling, starting from 1-benzyltetrahydroisoquinoline alkaloids. Most of the alkaloids of this type can be grouped according to the number, nature, and position of the bonds between the two halves, and according to the various combinations of substituents on the ring residues. The following fundamental types of bisbenzylisoquinoline alkaloids can be distiguished.

The simplest members are those in which the two halves are joined together by a tail-to-tail linkage.[1] An example of this type is provided by dauricine [10/19, from *Menispermum* sp. (MEN)]. In oxyacanthine [10/20, from species

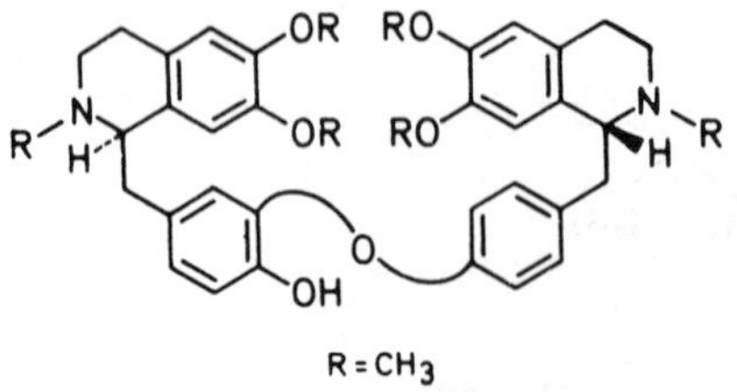

R = CH3

10/19, *Dauricine*

R = CH3

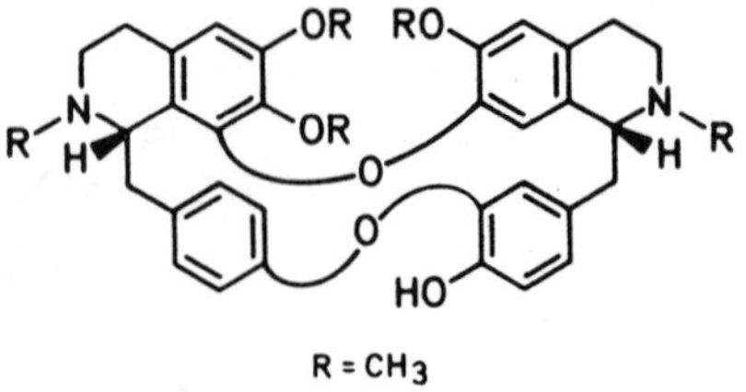

10/20, *Oxyacanthine*

---

[1]In a 1-benzylisoquinoline alkaloid the isoquinoline portion is referred to as the head and the benzyl portion as the tail.

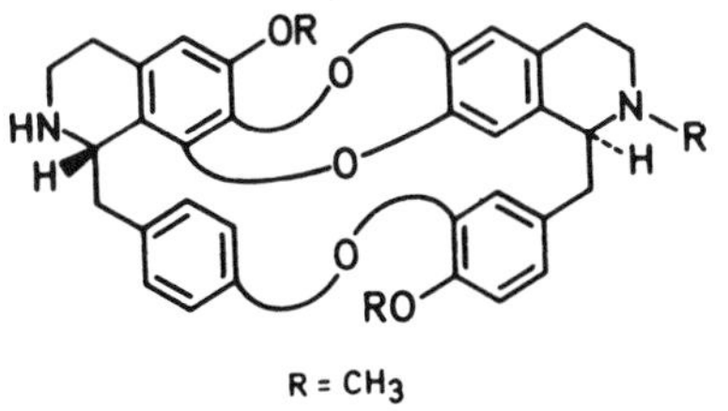

R = CH₃

10/21, *Trilobine* 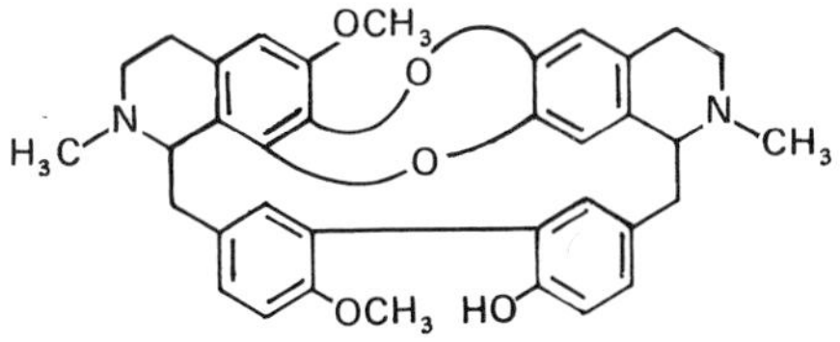   10/22, *Tiliacorine*

of the genera *Berberis* and *Mahonia* (BER), *Magnolia* and *Michelia* (MAG), and *Cocculus* (MEN)] the two halves are joined together by one head-to-head and one tail-to-tail bond. We can also have an extra bond between the heads of the two halves in alkaloids of the oxyacanthine type as, for instance, in trilobine [10/21, from *Cocculus* sp. (MEN)]. In addition to the diphenyl ether bonds between the two halves, there are also representatives that have C,C bonds, such as tiliacorine [10/22, from *Tiliacora acuminata* (MEN)].

In a third fundamental type of bisbenzylisoquinoline alkaloid the two halves are joined together by head-to-tail, tail-to-head bonds. A typical example is provided by tubocurarine [10/23, from *Chondrodendron* sp. (MEN) and tube curare arrow poison]. In this type there are also various sub groups with one, two, or three ether bonds between the building blocks. Tubocurarine is a muscle relaxant that interferes with depolarization, and as such has applications in surgery. However, it is inactive when administered orally.

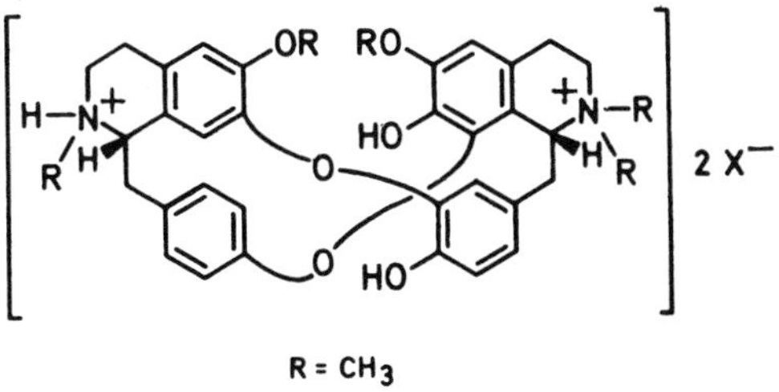

R = CH₃

10/23, (+)-*Tubocurarine*

A few bisbenzylisoquinoline bases have recently been produced synthetically by treatment of the corresponding monomeric alkaloid with potassium ferricyanide in basic medium.

Since the actual coupling mechanism is an oxidation, the bisbenzylisoquinoline alkaloids cannot be cleaved to give the monomeric basis under acid conditions. Reductive methods are required for this, and Na/liquid NH₃ has proved a very effective reagent for the purpose.

The alkaloid pilocerine [**10/24**, from *Lophocereus*, *Pachycereus*, and *Pilocereus* spp. (CAC)] belongs to the isoquinoline group, and has three basic units joined together by oxidative phenolic coupling. In contrast to the other alkaloids discussed in this section, the three units are not benzylisoquinoline alkaloids.

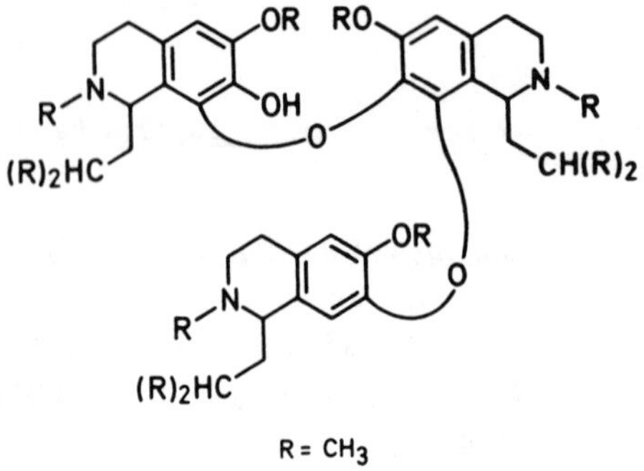

R = CH₃

10/24, *Pilocerine*

Oxidative coupling reactions are not necessarily confined to phenols. Two N_b-methyl tryptamine units can join together by oxidative coupling to give the bisindolenine **10/25** as in Scheme 23. The base **10/25** is equivalent to the tetraaminodialdehyde **10/26**. By interaction of the nucleophilic and electrophilic centers in **10/26** the following possibilities arise. Interaction of N_a and N_b with C-2′, and of $N_{a'}$ and $N_{b'}$ with C-2 leads to the quinoline derivative **10/27**. On the other hand if centers N_a and N_b are joined to C-2, while $N_{a'}$ and $N_{b'}$ are joined to C-2′ (the latter reactions can be better understood by reference to **10/25**), one obtains the bisindoline derivative **10/28**. Both **10/27** and **10/28** have been isolated from plants belonging to the Calycanthaceae: calycanthine (**10/27**) from

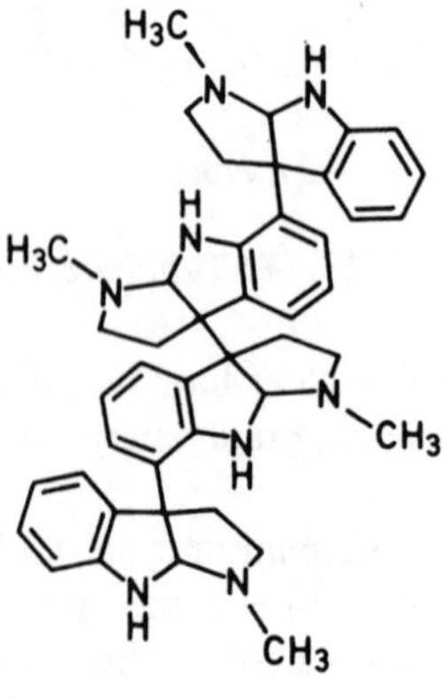

10/29, *Quadrigemine A*

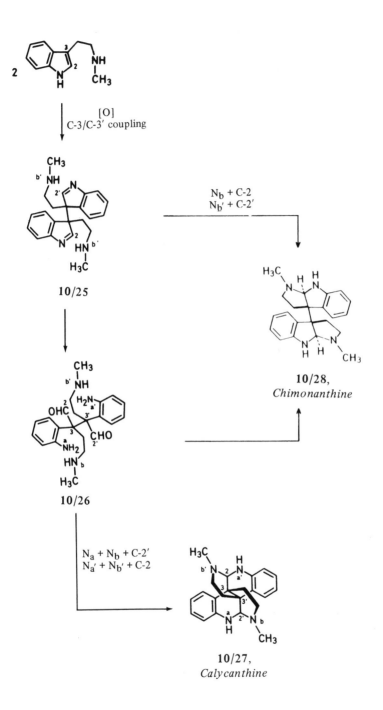

Scheme 23 Possible route for the formation of *Calycanthus* alkaloids.

Calycanthus and *Chimonanthus* spp. (CAL) as well as from *Bhesa* sp. (CEL), and chimonanthine (**10/28**) from *Chimonanthus* sp. Other naturally occurring derivatives of **10/27** are also known, and in addition there are some tri- and tetrameric examples belonging to this group such as quadrigemine A [**10/29**, from *Hodgkinsonia frutescens* (RUB)].

A radical coupling mechanism may also be involved in the formation of dithermamine [**10/30**, from *Thermopsis* sp. (LEG)], a dimeric alkaloid of the sparteine type.

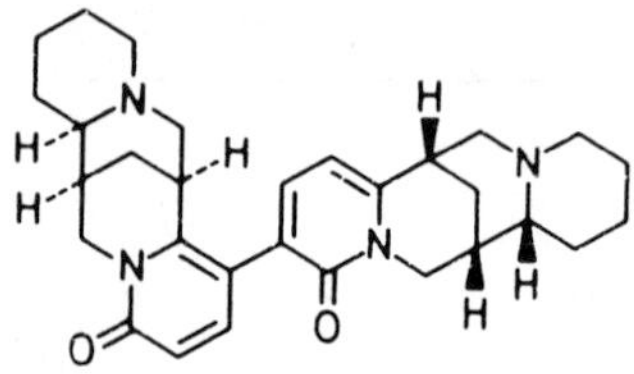

10/30, *Dithermamine*

10.5 LACTONIZATION

Dimerization of monomeric alkaloids is also possible through a double esterification as in the case of carpaine [**10/31**, from species of *Azima* (SAL), *Cerbera* (APO), and *Carica* (CAR)]. Only three representatives of this type are known.

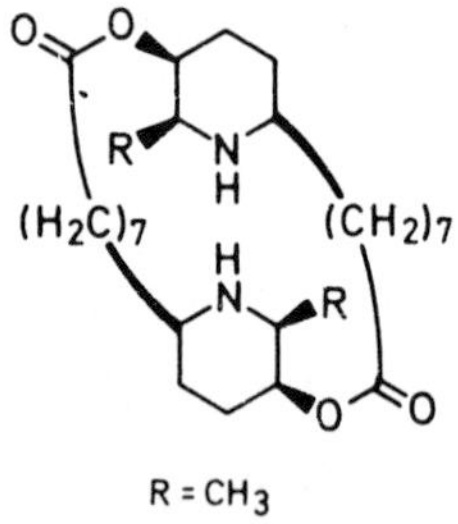

R = CH₃

10/31, (+)-*Carpaine*

10.6 COMBINED MECHANISMS

The alkaloid cancentrine [**10/32**, from *Dicentra canadensis* (FUM)] [R 10] is of special interest since it is evidently constructed from units of the morphine and cularine types. The hypothetical pathway shown in Scheme 24 throws a good deal of light on its biogenesis. This scheme is in accord with the biosynthesis of

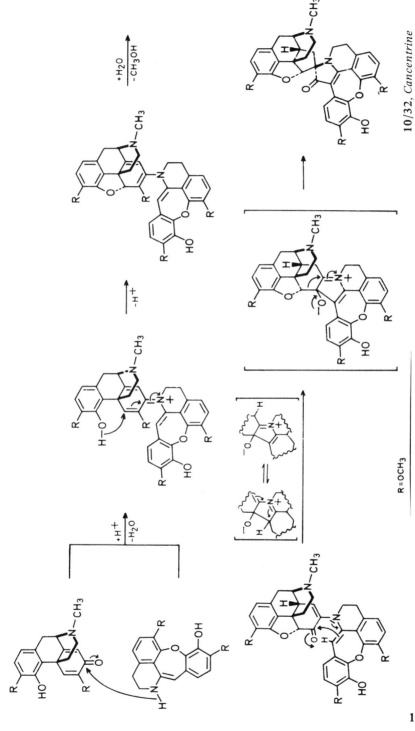

Scheme 24 Possible mode of biogenesis of cancentrine (10/32).

R = OCH₃

10/32, *Cancentrine*

173

the morphine alkaloids (see Section 6, Scheme 10) and gives a plausible explanation for the ring contraction that leads to the central five-membered ring.

Vinblastine (10/33) and vincristine (10/34), both from *Catharanthus roseus* (APO), are two bisindole alkaloids which are used intravenously against Hodgkin's disease (10/33) and against leukemia in infancy (10/34). So far these substances have been obtained only from natural sources, in which the concentration is extremely small.

10/33, *Vinblastine* R^1 = CH_3, R^2 = $COOCH_3$

10/34, *Vincristine* R^1 = CHO, R^2 = $COOCH_3$

In the synthesis of these bases in the plant from one unit of the *Iboga* type and one of the aspidospermidine type, it appears that a functional group in the *Iboga* unit has been split off, and this is responsible for the mode of coupling.

It is clear that many different mechanisms are involved in the coupling of monomeric units to form bisalkaloids, and in consequence there is no general method for cleaving these bases into the monomers again. Although in many cases acid conditions are effective, they are not applicable where oxidative or reductive changes around the positions of junction have taken place after dimerization, or where the coupling itself is primarily an oxidative reaction. The physical properties of these alkaloids are just as diverse as their structures. Thus a chromophore formed from two units which are coupled conjugatively gives a UV spectrum which is different from that obtained by summation of the spectra of the monomers. The complexities involved in dealing with physical properties and structural elucidation have been shown in the case of villalstonine (see Section 8), and other interesting examples could be added, but the scope of this book precludes a further treatment of this very interesting field related to the structural elucidation of alkaloids.

Examples of Alkaloid Synthesis

In a book dealing with aspects of alkaloid chemistry, a section on the important field of alkaloid synthesis is indispensible. The multiplicity of alkaloid structures is reflected in the profusion of synthetic methods, and it would be possible to compile an extensive work on this field alone. It is in consequence extraordinarily difficult to make a suitable choice from such a range when the number of examples must be limited. We have selected mesembrine to illustrate the synthesis of a simple tricyclic base, and the alkaloid porantherine to provide an example of a biomimetic synthesis. The construction of a macrocyclic ring is exemplified by the synthesis of the spermidine alkaloid oncinotine, and finally we shall discuss a stereoselective synthesis, that of vincamine. Although this selection might appear to be a rather modest one in view of the material available, the examples chosen should nevertheless provide an insight into the problems involved in the synthesis of alkaloids.

11.1 SYNTHESIS OF MESEMBRINE

Mesembrine (**11/1**) was first isolated in 1957 from *Mesembryanthemum tortuosum* (AIZ) [B 32], and the structure of (−)-mesembrine was elucidated in 1960

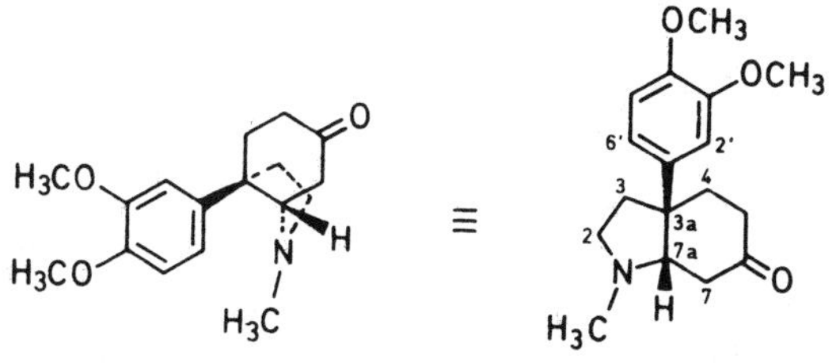

11/1, (−)-*Mesembrine*

175

[P 17]. The compound proved to have three rings, of which one is aromatic, another is heterocyclic, and the third is a cyclohexanone ring fused onto the heterocyclic (pyrrolidine) ring with a cis junction between the two. Special attention must be given to this latter steric arrangement in devising a synthesis.

The representation of mesembrine on the left indicates a similarity to the amaryllidaceous alkaloids of the ambelline type (see Section 5.1.7.17), while the right-hand structure gives a clearer picture of the relative dispositions of the three rings. Perhaps the first question to arise in connection with the synthesis of this alkaloid is how to construct the cyclohexanone ring. From an inspection of the right-hand formula it can be seen that the position of the keto group relative to the fused heterocyclic ring is the same as in 3-keto steroids. How is the keto group introduced into this position when synthesizing these steroids? A very effective method which is often used is to introduce it during the formation of the cyclohexanone ring. The so-called Robinson annelation reaction offers a method for this purpose.

In this procedure, which consists of a combined Michael and aldol-type reaction, cyclohexanone and butenone or derivatives thereof are reacted together in basic medium to form the required ring system:

11/2

Water is split off in the last step so that instead of the saturated ketone, the α,β-unsaturated derivative is formed. This type of reaction is thus not directly applicable to our problem—the synthesis of the corresponding ring in mesembrine—since on the one hand an α,β-unsaturated ketone is not required, and on the other hand the ring to be annelated is heterocyclic. If we take another look at the problem, this time putting in an amino nitrogen, we find that it must go in the α-position to the ring that has to be built on. Since the elimination of water is undesirable, instead of starting with a ketone it would be better to start with the corresponding alcohol. Thus the required starting material would be 11/3, in which C-2 has the oxidation number of an aldehyde. Apart from the

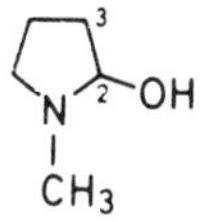

11/3 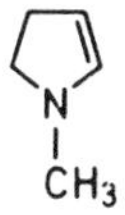   11/4

fact that **11/3** loses water very readily to form the enamine **11/4**, position 3 of **11/3** is not sufficiently activated to permit abstraction of a proton with base so as to form a stage corresponding to the enolate ion above. On the other hand, **11/4** would react with butenone because of its enamine structure:

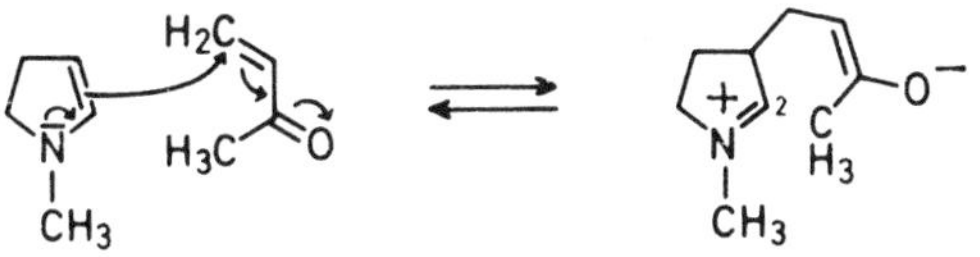

11/5

The intermediate **11/5** formed from the enamine **11/4**, and **11/2** formed from cyclohexanone correspond to one another. The electrophilic carbon atom that is essential for the ring closure forms part of the ketone group in **11/2**, while in **11/5** it is C-2. The subsequent reaction of **11/5** would thus lead to an analogous ring closure to **11/2**, but without the subsequent elimination of water, which is important for the solution of our problem.

11/5   11/6

The compound **11/6** exhibits all the necessary structural elements present in the lower portion of the molecule of mesembrine. Thus the first part of the problem is solved, and we can turn our attention to the next question: how can we synthesize the enamine **11/7**, which we need as an intermediate stage? This will depend on methods for forming dihydropyrroles, which can for instance be

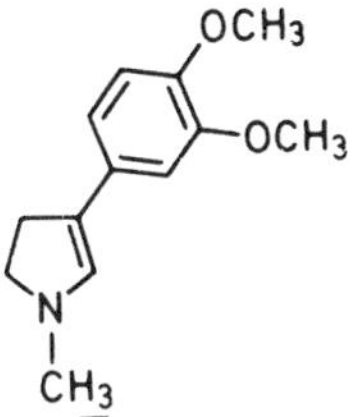

11/7

obtained by splitting off water from 2- or 3-hydroxypyrrolidines. We have already noticed the tendency for elimination of water from 2-hydroxypyrroli- dines. The compound **11/3** is equivalent to 4-methylaminobutanal, and its 3-hydroxy analogue is accessible by reduction of 1-methyl-3-pyrrolidone. A formal reduction takes place in fact when 1-methyl-3-pyrrolidone is substituted at position 3 by a nucleophile:

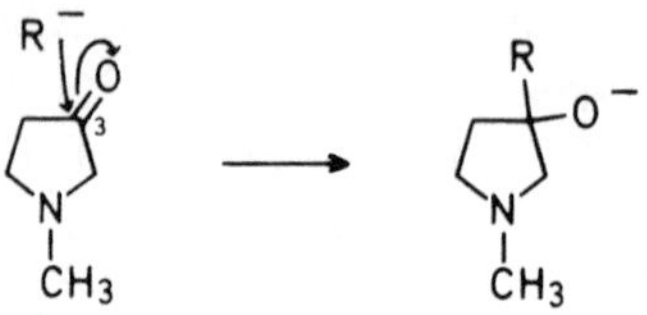

The choice of the R group is obvious. It must consist of the remaining aroma- tic ring, correctly substituted. In order to carry out a nucleophilic reaction with an aromatic ring, organometal reagents are necessary, such as Li- or Mg-phenyl derivatives. The general outline of a synthesis for mesembrine can now be sketched out (Scheme 25). The synthesis was in fact carried out by this method [C 15], and an overall yield of 37.4% of (±)-mesembrine was obtained. The cis- arrangement in the reaction product is automatically assured. The stage between **11/7** and **11/1** is **11/8**, in which the attack of the nucleophilic center on C-7a takes place trans to the phenyl nucleus, as required sterically. Consequently the hydrogen atom at C-7a and the phenyl nucleus are cis to one another as in the natural product. The formation of the trans isomer is thus prevented since a maximal overlap of the π orbitals is not possible in that case. The intermediate **11/7** has also been synthesized by another method (see [K 12], [S 47]).

11.2   A BIOMIMETIC SYNTHESIS: PORANTHERINE

Poranthera corymbosa Brogn. (EUP) is endemic in Australia and contains a num- ber of alkaloids. The total yield calculated on the dry weight of plant material amounts to 0.4%. The following six alkaloids can be obtained from it in reason- able amounts:

Porantherine (**11/9**)

Porantheridine (**11/10**)

Poranthericine (**11/11**)

Porantherilidine (**11/12**)

Porantheriline (**11/13**)

O-Acetylporanthericine (**11/14**) [J 7]

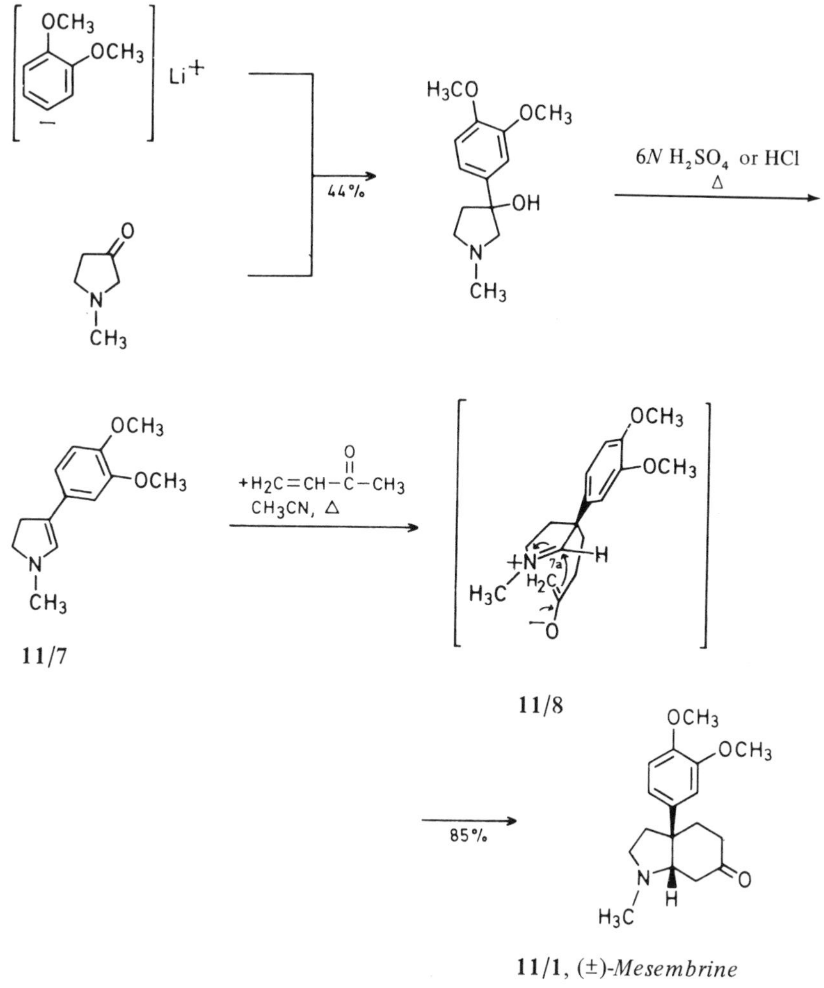

Scheme 25 Synthesis of (±)mesembrine (11/1) [C15].

The structures of these bases were determined by X-ray crystallographic analysis (11/9: [D 5], [D 6] ; 11/10: [D 7], [D 8]; 11/11: [D 7], [D 9] ; 11/12: [D 10] ; the structures of the remaining two bases were deduced by chemical correlation with the main alkaloid) [J 7] .

The nucleus of porantherine (11/9)—9b-azaphenalene—was previously unknown in a natural plant product, but alkaloids with the same nucleus such as propyleine (11/15, from *Propylaea quatuordecimpunctata* L.) [T 16] have been isolated from beetles of the family Coccinellidae. Other substances with similar

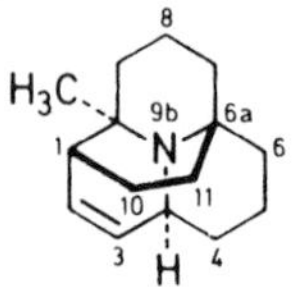

11/9, *Porantherine*[1] 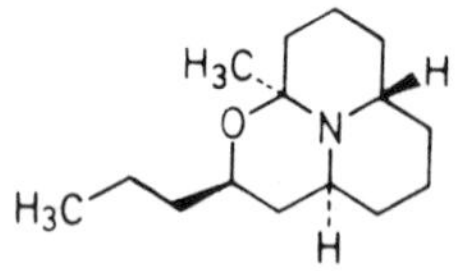   11/10, *Porantheridine*   11/11, *Poranthericine* R = H

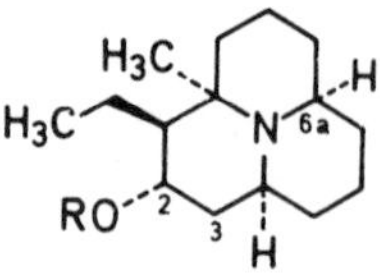

11/14, *O-Acetylporanthericine*
R = COCH₃

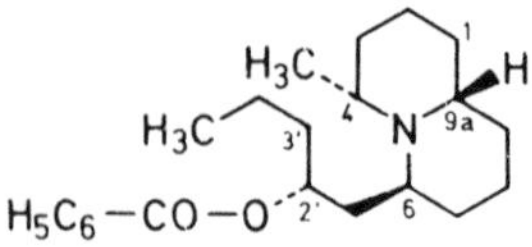

11/12, *Porantherilidine* 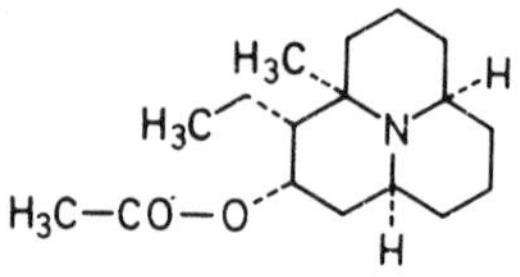   11/13, *Porantheriline*

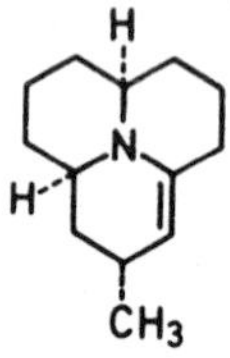

11/15, *Propyleine*

structures include coccinelline from *Coccinella septempunctata* L. [T 17] as well as hippodamine and convergine from *Hippodamia convergens* [T 21] .

The biosynthetic relationships of the *Poranthera* alkaloids 11/9-11/14 are self-evident. Porantherine (11/9) is probably formed by ring closure between a suitably functionalized methyl group in the ethyl residue of poranthericine (11/11) and position C-6a, accompanied by loss of water. Poranthericine (11/11) in turn must be derived from porantherilidine (11/12) by ring closure between C-4 and C-3', for which suitable functional groups in the relevant positions would likewise be required. It should be noted that C-9a in 11/12, which corresponds to C-6a in 11/11, is epimeric. The formation of the tetrahydrooxazine ring in 11/10 is explained by a ring closure between C-2'−OR and C-4 of 11/12.

If a retrosynthetic (antithetic) analysis of porantherine (11/9) is carried out along these lines, a feasible synthetic route such as that of Scheme 26 is arrived at. The point of departure for this scheme is the 2,3 double bond in 11/9, which could be formed by elimination of water from a C-2−OH, or possibly a C-3−OH

---

[1]The numbering system given in [D 7] differs from that shown in 11/9.

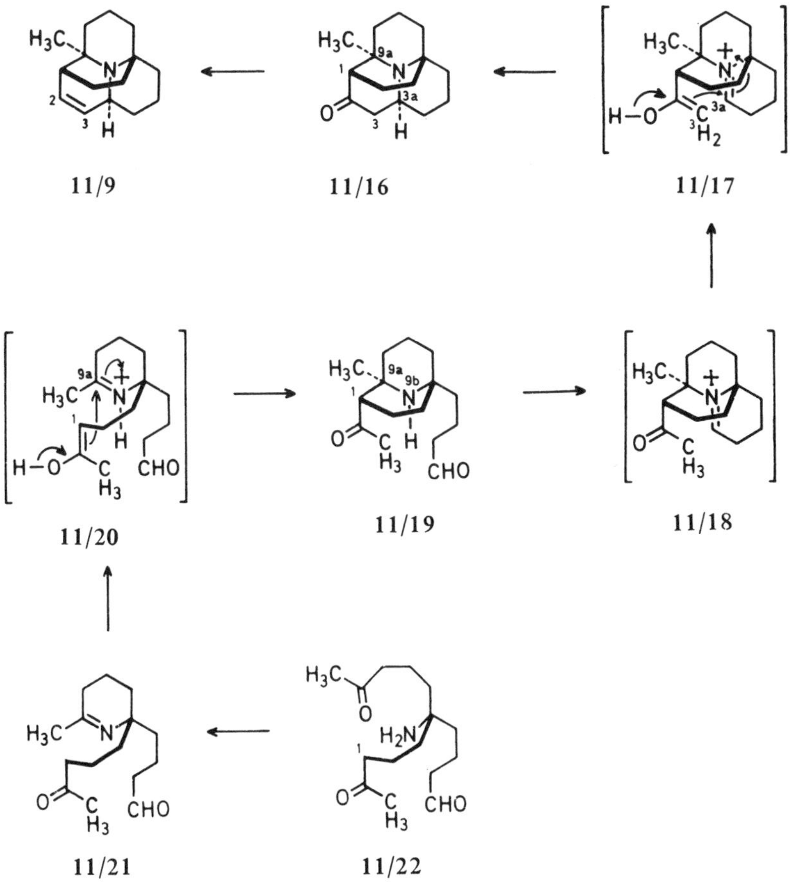

Scheme 26   Retrosynthetic analysis of porantherine (11/9).[2]

group. A C-2—OH group could probably be formed by reduction of the C-2 keto compound 11/16, in which the keto group has a twofold β relationship with respect to the amino nitrogen (via 1,9a and via 3,3a). A structure of the type:

$$\text{>N--CH}_2\text{--CH}_2\text{--}\overset{\displaystyle O}{\overset{\displaystyle \|}{\text{C}}}\text{--R}$$

is formed by a Mannich reaction between an amine, an aldehyde or ketone, and a ketone with an α proton:

[2]The system of numbering for porantherine is used in Scheme 26 and in the discussion on its synthesis.

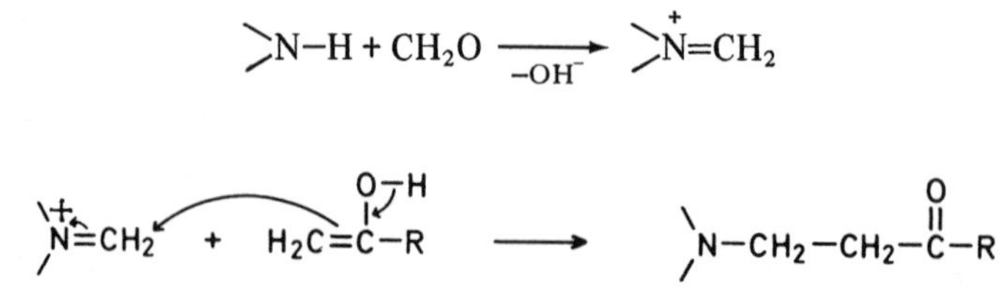

$$>N-H + CH_2O \xrightarrow[-OH^-]{} >\overset{+}{N}=CH_2$$

$$\overset{\backslash}{\underset{/}{N}}\overset{+}{=}CH_2 + H_2C=\overset{\overset{O_{\overline{}}H}{\|}}{C}-R \longrightarrow \overset{\backslash}{\underset{/}{N}}-CH_2-CH_2-\overset{\overset{O}{\|}}{C}-R$$

Thus **11/16** is a Mannich base, which can be formed from **11/18** via **11/17** by condensation between C-3 and C-3a; **11/18** is formed from the aldehyde **11/19** by elimination of water. In **11/19** there is another keto group in a β-position to an amino nitrogen, as in the case of **11/16**. A bond between C-1, C-9a, and N-9b can again be formed by a Mannich reaction, for which the open-chain compound **11/22** serves as starting material. The sequence set out in Scheme 26 achieves the transformation of **11/22** into **11/9** simply by the use of Mannich reactions followed by loss of water, plus a reduction.

It is unlikely that the compound **11/22**, which has an aldehyde and two keto groups, would form **11/16** directly. There are too many other possibilities of reaction. For a laboratory synthesis the carbonyl groups of **11/22** must be protected, and the reaction must proceed in stages.

The compound **11/22** is symmetrical,[2] likewise its analogue **11/27** with protected carbonyl groups, which can be synthesized as follows. Two equivalents

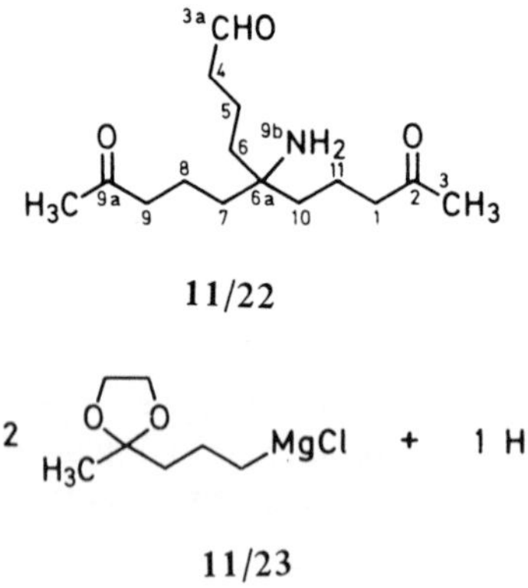

11/22

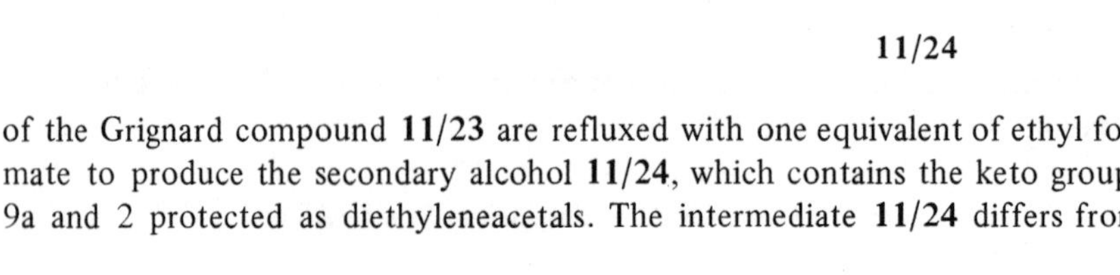

11/24

of the Grignard compound **11/23** are refluxed with one equivalent of ethyl formate to produce the secondary alcohol **11/24**, which contains the keto groups 9a and 2 protected as diethyleneacetals. The intermediate **11/24** differs from

11/22 in the nature of the substituents at C-6a, to which an amino nitrogen as well as the butanal chain must be attached. The oxidation number at C-6a is +I, which is too low for this purpose, so **11/25** is oxidized with 6 equivalents of Collins reagent at 20°C in dichloromethane to the ketone **11/25** (in 93% yield). The nitrogen can now be attached, and this is done in a sealed tube with methyl-amine in toluene, using a molecular sieve (4° A) to remove the water that is produced and thus promote the formation of **11/26**. The center 6a is electro-philic in **11/26**, and the alkyl chain can be attached by a nucleophilic substitution reaction with 1-pent-4-enyllithium, which is prepared from 5-bromo-1-pentene by treatment with lithium in ether. The compound **11/27** constitutes a derivative of **11/22** with all four functional groups protected. By careful treatment of **11/27** with 10% hydrochloric acid, the two keto groups

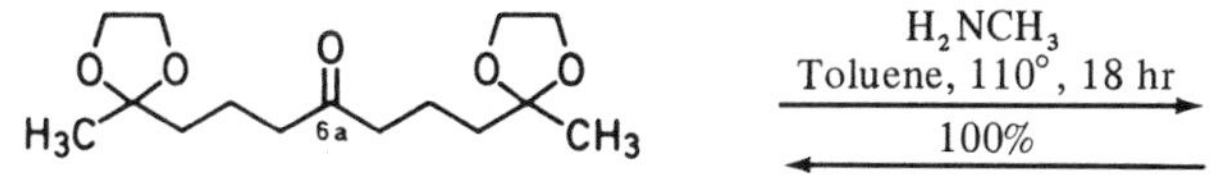

11/25

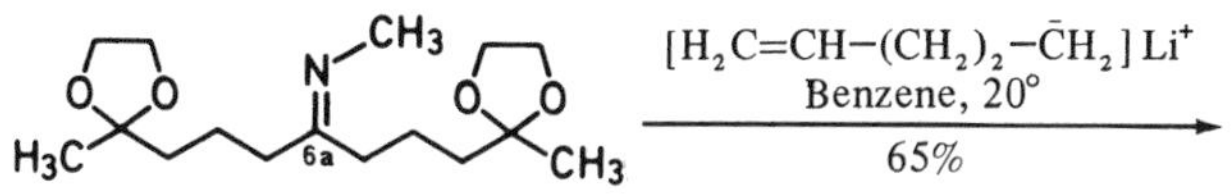

11/26

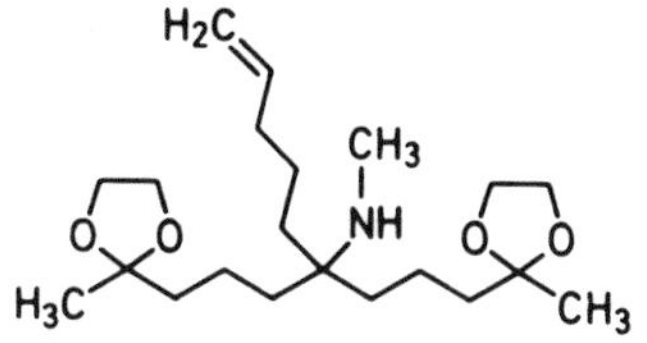

11/27

are released and one of them simultaneously reacts with the amino group to form the enamine **11/28**. The latter corresponds to **11/21**, which forms a stage in Scheme 26. The bicyclic compound **11/29** is obtained by a further Mannich reaction in the presence of *p*-toluenesulfonic acid and isopropenyl acetate. Under the reaction conditions used, the C-2 keto group is first transformed into the corresponding enol acetate which then reacts to give **11/29**. The configura-tion at C-1 of the latter is undetermined, and in any case there is a possibility of epimerization in the subsequent reactions.

The next stage to aim at is a ring closure between the nitrogen and an alde-hyde function at C-3a. This means that an aldehyde group must be formed at that position, and the methyl group must be removed from the nitrogen. The methylimino group that protects the nitrogen can be oxidized with Collins reagent (CrO$_3$-pyridine, 10 eq. in CH$_2$Cl$_2$ for 72 hours; 11/29 $\longrightarrow$ 11/30) to an N-formyl group, which can be easily removed hydrolytically at a later stage. To convert the vinyl double bond into the C-3a aldehyde group, osmium(VIII) oxide and sodium metaperiodate in aqueous tertiary butanol are used. The

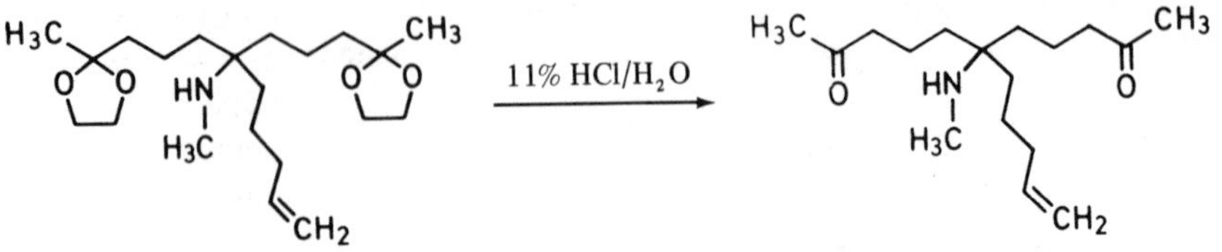

11/27

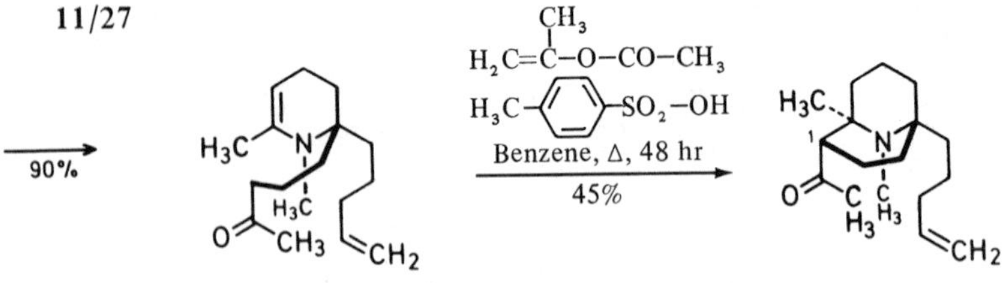

11/28

11/29

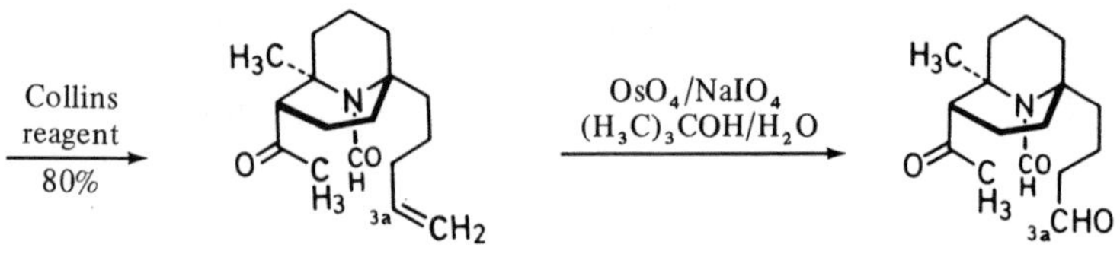

11/30

11/31

ketoaldehyde 11/31 is immediately converted into the acetal 11/32 with an excess of glycol in benzene. The protection of the carbonyl groups is necessary to avoid condensation products in the subsequent base-catalyzed hydrolysis with 3N KOH. The aminoacetal 11/33 so formed is converted with 10% hydrochloric

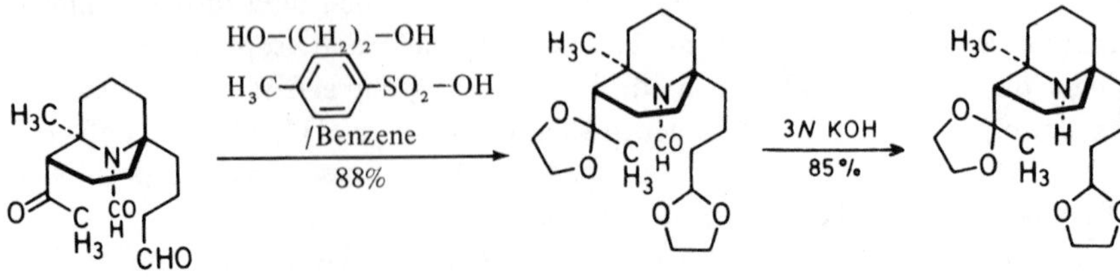

11/31

11/32

11/33

acid into the unstable tricyclic enamine **11/34**, of which the immonium form is **11/18** (see Scheme 26). The final cyclization step consists in the formation of **11/35** from **11/34**, and is carried out by boiling with *p*-toluenesulfonic acid under reflux. Borohydride reduction of the keto group affords the alcohol, which can be dehydrated with thionyl chloride and pyridine to give porantherine (**11/9**) in an overall yield of ca. 2.5%. The synthetic (±) porantherine was identified by comparison of its physical properties with those of the natural product [C 16].

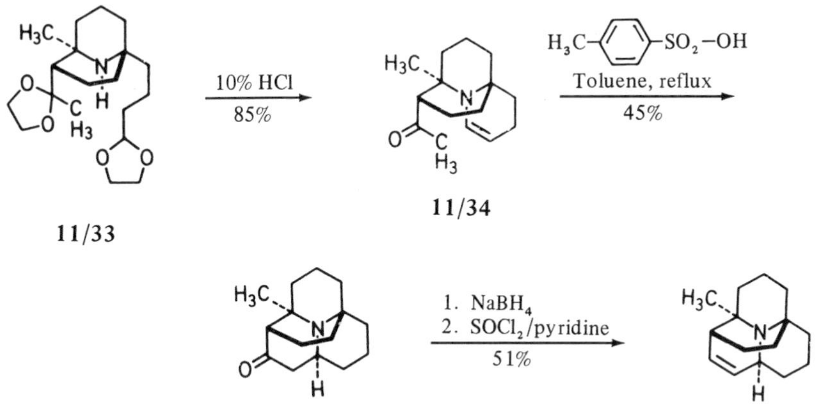

11.3   SYNTHESIS OF THE MACROCYCLIC SPERMIDINE ALKALOID ONCINOTINE

Oncinotine (**11/36**), the main alkaloid of *Oncinotis nitida* (APO), belongs to the class of spermidine alkaloids since its basic component is the biogenic amine spermidine (see Section 5.3). The other part of the molecule is derived from a

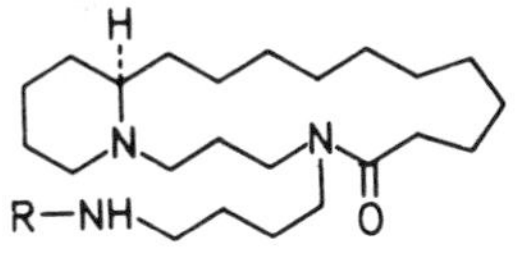

11/36, R = H Oncinotine
11/37, R = COCH₃ *N*-Acetyloncinotine

C_{16}-carboxylic acid. Oncinotine is an oil that has not yet been crystallized. Two isomeric compounds occur along with it in the plant: neooncinotine (11/38), in which the spermidine unit is incorporated into the molecule in the reverse fashion to 11/36, and isooncinotine (11/39), with the same spermidine unit as 11/38 built into the molecule, but in a larger lactam ring. From the crude alkaloid mixture obtained from the plant, the base 11/39 is easily separated from

11/38, *Neooncinotine*   11/39, *Isooncinotine*

11/38 and 11/36, but it is not possible to separate the latter two directly since their properties are too similar. Their derivatives can, however, be separated chromatographically [G 17].

In planning a synthesis of the main alkaloid oncinotine (11/36), the formation of the 17-membered lactam ring would appear to be an important step. For this purpose it might be possible to cyclize a previous stage 11/41 to produce 11/40 by the choice of a suitable leaving group X in 11/41. But before embarking on a synthesis of this kind there are certain problems that one should be

11/40   11/41

aware of and which will have to be solved. In a critical stage of the reaction it will be necessary to close a large ring. In contrast to the normal size rings with five to seven members, the formation of a so-called large ring is not specially favored at normal concentrations. Molecules such as 11/41 tend to dimerize or polymerize instead. However, if reaction conditions are chosen whereby 11/41 is kept in low concentration then, in terms of the Ziegler dilution principle [Z 3], cyclization should be preferred over dimerization or polymerization. A further problem arises with the choice of the nitrogen substituent R. A reasonable choice would be either R = H or R = $(CH_2)_4-NH_2$. If R = H, the ring closure can take place only in the way shown, since a quaternization onto the piperidine nitrogen is excluded, but then the tetramethyleneamino group that is present in the alkaloid must be joined to the secondary amido nitrogen of 11/40 in a separate synthetic step [S 48]. If on the other hand R = $(CH_2)_4-NH_2$, unless some preventive measure is taken, the secondary and primary amino groups could both take part in the cyclization reaction to give not only oncinotine (11/36) but an isomer, pseudooncinotine (11/42), which so far has not been found to occur in nature.

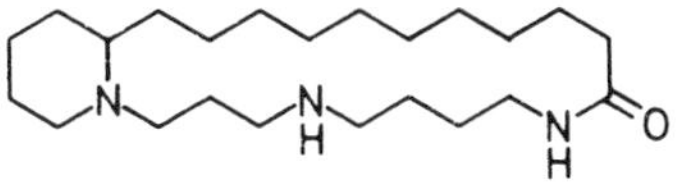

11/42, *Pseudooncinotine*

If we make R = (CH₂)₄–NH₂, the formation of **11/42** can be prevented by the introduction of a protective group on the primary amino nitrogen. The protective group must be so chosen that when it is removed, the lactam group is not opened again; in other words, acid and basic reaction conditions must be avoided. Thus if one chooses an acetyl group to protect the primary amino function at the end of the chain, then acid conditions (conc. aqueous HCl, sealed tube, 120°C, 3.5 hours) must be used to remove it from the *N*-acetyloncinotine (11/37) that is formed. Under these conditions the yield of oncinotine (11/36) is only 55%, the most important byproduct being the diaminocarboxylic acid **11/43**.

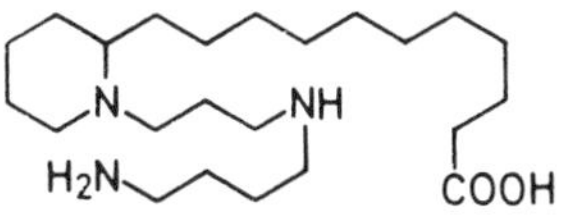

11/43

Thus in choosing the protective group, its selective removal must be kept in mind. The benzylamino group seems to meet the necessary conditions in full since its removal can readily be achieved by hydrogenolysis, which would not affect any other group in the molecule. Thus **11/44** can be quantitatively converted into oncinotine (11/36). The ring closure reaction can with advantage

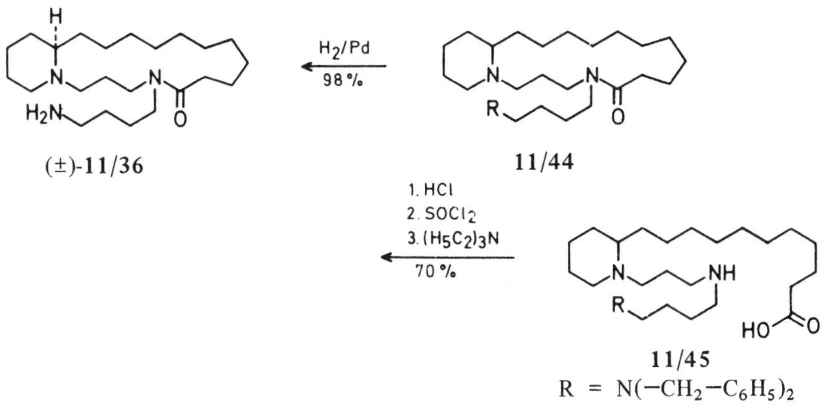

(±)-11/36

H₂/Pd
98%

11/44

1. HCl
2. SOCl₂
3. (H₅C₂)₃N
70%

11/45

R = N(–CH₂–C₆H₅)₂

be carried out with the acid chloride of the amino acid **11/45**, which is first converted into its trihydrochloride, then the dried salt is transformed into the acid chloride with thionyl chloride. The cyclization is carried out according to the Ziegler dilution principle in ca. 6×10^{-7} molar benzene solution at $20°C$ in the presence of triethylamine as base, which results in a 70% yield of **11/44**.

The compound **11/45** is a dialkyl piperidine derivative which is substituted on the nitrogen and also in the α-position. An *N*-alkylation reaction applied to the α-alkylpiperidine **11/46** can be employed in synthesizing this intermediate. It is advantageous to use the ester **11/46** rather than the corresponding amino acid.

11/46

The next question to arise, which alkyl residue to use for the *N*-alkylation, is similar to that discussed previously in connection with the overall strategy of oncinotine synthesis. The reaction **11/46** ⟶ **11/45** requires a 1,6-diaza-nonyl residue, which can be attached as such directly to the piperidine derivative **11/46**. Alternatively it would be possible to attach part of this group initially and then extend it to the required length.

In cases like this the most rational approach is to try as far as possible to link the individual units together close to the end of the synthetic sequence. This general observation is of fundamental importance for a synthesis. The overall yield of the final product is higher if one first prepares two intermediates separately from one another, which are then linked together at as late a stage as possible, instead of carrying out the synthesis in a series of consecutive steps, even though the total number of individual steps and the yield from each one may be the same. The price of the product is a direct function of the overall yield, and it comprises not only the cost of the chemicals themselves but also the expenditure of time. In line with these considerations it is essentially more profitable to synthesize two separate portions of the final molecule that are as large as possible, and then join them together (see [I 1]).

From this discussion the first alternative, addition of the 1,6-diazanonanyl residue to **11/46**, is evidently to be preferred. The alkyl residue required can be represented by the general formula **11/47**, and consists of a putrescine derivative with a trimethylene chain attached to the nitrogen atom. The residue X must be

11/47, general structure for the
synthetic unit

a leaving group which can take part in the alkylation of **11/46**. Suitable groups include the halogens Br and I, and Br is preferred in this case. The general formula for the synthetic unit required is then **11/47a**, and to avoid inter- and intramolecular N-alkylation, both amino nitrogens of **11/47a** must bear a protective

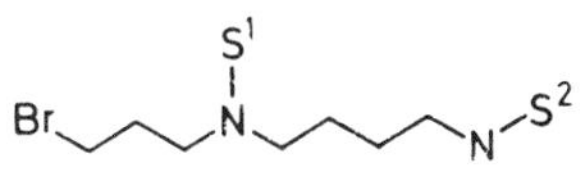

11/47a

group. The properties of these protective groups S^1 and S^2 must be such that the nitrogen atoms are no longer basic, and thus the N,N-dibenzylamino group that was used in the reaction **11/45** ⟶ **11/44** is not applicable here. Furthermore, S^1 and S^2 must differ from one another owing to the structure of the intermediate **11/45**, and each must be selectively removable. This latter requirement is not only necessary for the synthesis of oncinotine, but it is also connected with the desire to extend the method to make accessible the isomeric pseudooncinotine (**11/42**) through the same reaction sequence.

The commercially available putrescine (**11/48**) is an obvious starting material for the intermediate **11/47a**, but it involves another requirement for the protective group S^1: it must permit the protected putrescine to be N-alkylated in order to introduce the trimethylene residue. How might this be achieved?

Putrescine (**11/48**) is first treated with one equivalent of acetic anhydride in aqueous potassium carbonate, and the product is converted to **11/49** with one equivalent of p-toluenesulfonyl chloride (TosCl). This secondary sulfonamide, when treated with a strong base (NaH), is converted into the anion, which reacts with 1,3-dibromopropane to give **11/50**. In order to avoid a double addition to

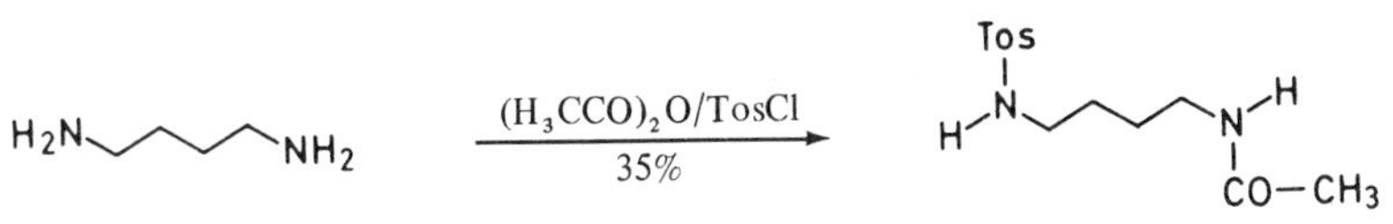

11/48

11/49

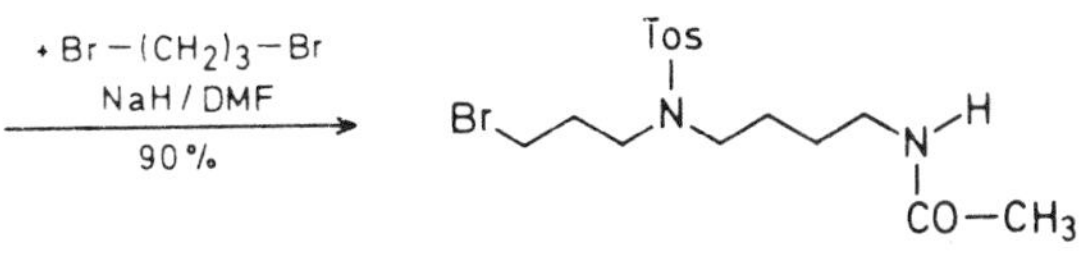

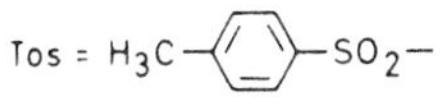

11/50

the dibromo compound, a large excess (ca. hundredfold) of 1,3-dibromopropane is used. The *N*-alkylation reagent **11/50**, which has the general formula **11/47a**, can now be reacted with **11/46**. The reaction is carried out in ethanol solution in the presence of NaI (to exchange Br for I) and a base as proton acceptor, and affords an 88% yield of **11/53**. The conversion of **11/53** into **11/45** requires the exchange of the acetyl for two benzyl groups by acid hydrolysis followed by reaction with benzyl chloride; the removal of the ester group, which occurs during the acid hydrolysis; and the selective cleavage of the tosyl residue for which electrolysis has proved especially suitable.

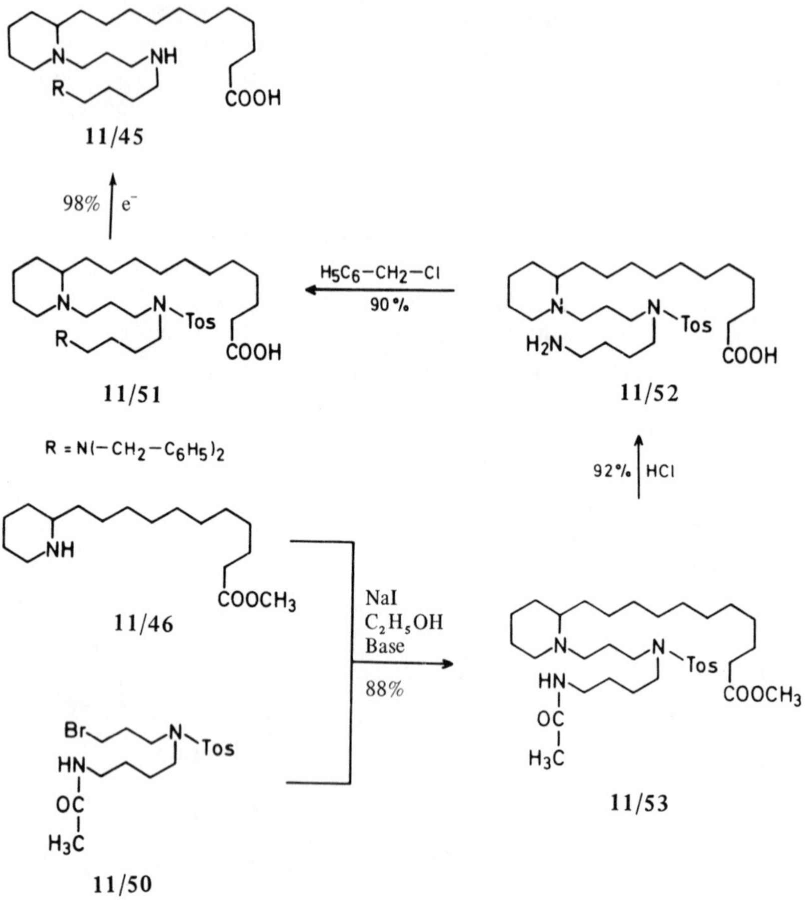

The synthesis of the piperidine intermediate **11/46** was carried out according to Scheme 27.

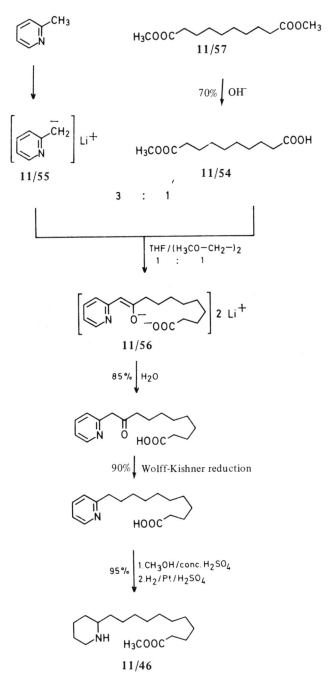

Scheme 27 Synthesis of the oncinotine intermediate **11/46**.

Among the various reactions given in Scheme 27, attention should be called to the interaction of the hemiester **11/54** with the lithium compound **11/55**, in which the main product formed is **11/56** instead of the corresponding ester, as might be expected. Usually a carboxylic acid reacts in the proportion of 1:2 with an organolithium compound to form the enolized ketone.

The overall yield of (±)-oncinotine (**11/36**) amounts to 25% when calculated on the basis of dimethyl sebacate (**11/57**), and 15% when calculated on putrescine. It should be observed, however, that the poorest yields are encountered at the beginning of the synthetic sequence, while the last stages involving expensive reagents proceed in 50% yield [G 18]. Syntheses of neooncinotine (**11/38**), isooncinotine (**11/39**), and pseudooncinotine (**11/43**), all isomeric with oncinotine, are readily possible by modification of the synthetic intermediate **11/50** [G 18].

11.4 STEREOSELECTIVE SYNTHESIS: VINCAMINE

Vincamine (**11/58**) is an indole alkaloid that has been isolated from various *Vinca* spp. (APO) (see Section 5.1.2.3.3). It has the absolute configuration shown in **11/58** (3*S*, 14*S*, 16*S*). The value found for the specific rotation is dependent on the solvent—thus $[\alpha]_D = -6°$ in $CHCl_3$ and $+41°$ in pyridine—and we must therefore omit the usual sign prefixed to the name. The alkaloid is of increasing pharmacological interest as a regulator of cerebral blood circulation.

11/58, *Vincamine*[3]

Vincamine has a great attraction for the synthetic chemist, as shown by the large number of publications on nonstereoselective [G 20] [K 15] and stereoselective synthesis [H 25], [H 26], [S 57], [T 22], [W 18], [P 19]. We shall discuss in detail the most recent synthesis [P 19], which proceeds with an overall yield of 4.5% if we neglect the resolution of racemates, and as a preliminary we shall first consider various synthetic aspects related to the vincamine molecule.

Of the three chiral centers of vincamine (**11/58**) at C-3, C-14, and C-15, the center 14 is located α to the indolic nitrogen and has an oxidation number of

---

[3]To insure a uniform overall method of presentation, all the synthetic units are numbered in accordance with the vincamine system of numbering.

+II, the same as for a ketone. The tautomer **11/59** would be expected to form an equilibrium with vincamine, although it should be noted that the equilibrium lies in the direction of the cyclic form. In addition to vincamine (**11/58**), a second compound, 14-epivincamine (**11/60**), should also be capable of existence; in fact, **11/60** also occurs naturally in the evergreen plant *Vinca minor*.

11/58   **11/59**   **11/60**

The ratio of **11/58** to **11/60** occurring in the equilibrium mixture is dependent on the stereochemistry of the rest of the molecule, in particular on the configurations of the centers 3 and 16, but we need not go into these details for the moment. Since the ketoamine **11/59** is not stable, a protective group for the keto function must be introduced in order to obtain this open-chain form rather than the alkaloids **11/58** and **11/60**. An enol ether suggests itself for this purpose, and the compound **11/61** should be stable with ring D opened.

11/61

It is easier to seek a method of synthesizing **11/61** instead of **11/58** or **11/60** since the number of rings has been reduced by one, likewise the number of chiral centers. An examination of the ring system in **11/61** shows a similarity to the tetrahydro-β-carboline system (**11/62**), or more simply to the tetrahydroisoquinoline **11/63**.

Both can be prepared by the same synthetic method. On treatment of an open-chain *N*-acetyl derivative such as **11/66** or **11/67** with a dehydrating agent like P_2O_5 or $POCl_3$ in benzene or tetralin, a ring closure of the Bischler-Napieralski type occurs with formation of the dehydro derivative **11/64** or **11/65**. A subsequent reduction carried out with $NaBH_4$ proceeds smoothly and affords racemic **11/62** or **11/63**.

If one uses this synthetic sequence for the vincamine intermediate **11/61**, the starting material required is **11/68**, and the product of the Bischler-Napieralski reaction is subsequently reduced. With the introduction of **11/68** a further

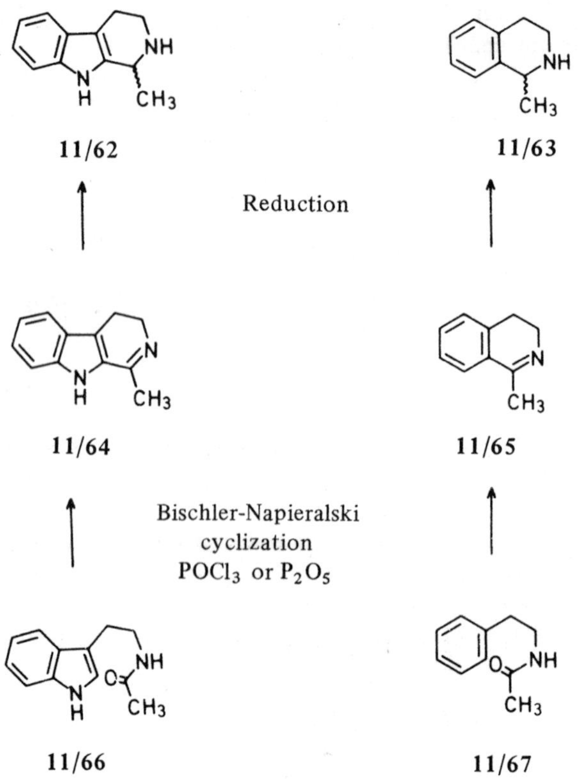

11/62   11/63

↑   Reduction   ↑

11/64   11/65

↑   Bischler-Napieralski   ↑
              cyclization
            POCl₃ or P₂O₅

11/66   11/67

simplification of the problem is achieved as compared to **11/61**, since the number of rings and chiral centers has been reduced to one. Moreover the center C-16 of **11/68** must determine the chirality of C-3 in **11/61**. In this connection it is important to note that C-16, the only chiral center in **11/68**, cannot be epimerized by the adjacent lactam group at C-3 since C-16 is a quaternary carbon atom, and as a result there is no problem about opening the lactam to the corresponding carboxylic acid and secondary amine. Pursuing the "analytical" approach to the development of the synthesis, we may consider the ester

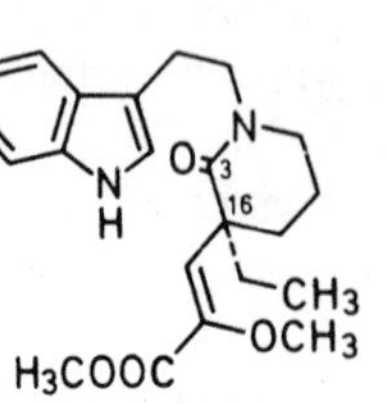

11/68 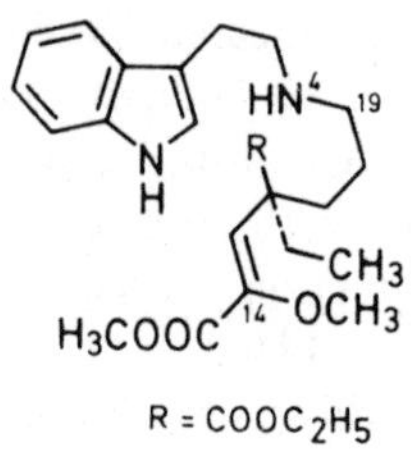   11/69

amine **11/69** as a possible precursor for **11/68** in order to avoid the preparative difficulties involved with an amino acid. The commercially available tryptamine **11/70** is an obvious choice as one component for preparing **11/69**, and at the same time this suggests the kind of bond required between N-4 and C-19, which could be formed from **11/70** by an S_N2 reaction with an aliphatic precursor bearing a suitable leaving group, or alternatively by reduction of a Schiff base formed by reaction of **11/70** with an aliphatic aldehyde. We need not consider the second possibility further, since the synthesis of an aldehyde of this kind involves unnecessary problems. There is already one protected carbonyl group required for C-14, and a similarly protected aldehyde group for the C-19 carbon would cause certain difficulties. We shall therefore turn our attention to the first alternative, alkylation, for the formation of the N-4—C-19 bond. The aliphatic component required for this is **11/71**, in which X is a suitable leaving group. Compound **11/71** contains a chiral center that is crucial for the synthesis of

11/70

R = COOC$_2$H$_5$

11/71

vincamine (**11/68**). The intermediate **11/71** must also be synthesized since it is not available commercially: how could it be made? The C-14=C-15 double bond could be formed from the aldehyde **11/73** and methoxytrimethylphosphoroacetate (**11/72**) by a Wittig-Horner reaction. There would be no risk of epimerization of chiral center 16 in **11/73** for the same reason as before—it has a quaternary carbon next to the carbonyl groups.

11/72

11/73

The synthesis of the reactive side chain that comprises carbon atoms 17, 18, and 19 as well as X would appear to be best carried out in two stages. If one takes advantage of the acidity of the methine proton attached to C-16 in **11/74**, then a molecule such as

$$\overset{17}{Y}-\overset{}{CH_2}-\overset{18}{CH_2}-\overset{19}{CH_2}-X$$

where X and Y are leaving groups, should react quite well in the proposed *C*-alkylation with loss of Y from C-17. It could also lose X from C-19. We must in consequence consider the ambidence of this compound. One possibility is an allyl derivative with a C-18=C-19 double bond instead of X, which can be functionalized later by insertion of X in an anti-Markovnikov reaction after Y has been removed in the *C*-alkylation reaction. Moreover, the primary alkylation product **11/75** is a suitable stage for resolution, since the epimerization of center 16 is no longer possible.

H_5C_2OOC

OHC⁻¹⁶⁻H

CH₃

11/74

H_5C_2OOC CH₂

OHC⁻¹⁶⁻‖

CH₃

11/75

This synthetic pathway, which has been developed by a retrospective approach, still has one unsatisfactory feature which could easily lead to complications. The conversion of compound **11/71** into **11/68** requires a hydrolysis which, if carried out under acid conditions, would most probably result in simultaneous cleavage of the other ester group and hydrolysis of the enol ether. It would therefore be more convenient to carry out the Wittig-Horner reaction at a stage where C-3 has already been lactamized.

The synthesis of vincamine (**11/58**) is set out in Scheme 28 [P 19]. The starting material is racemic ethyl 2-formylbutanoate (**11/74**), which gives a mixture of *O*- and *C*-alkylated products on treatment with the Hünig base $C_2H_5N[CH(CH_3)_2]_2$ and allyl bromide. On heating, this mixture is converted into (±)-**11/75** by a Claisen rearrangement. Acetalation of the aldehyde group affords (±)-**11/76**, which is converted by alkaline hydrolysis into the carboxylic acid (±)-**11/77**. This acid provides a favorable opportunity for the first resolution of the synthesis, using the optically active base L-(+)-pseudoephedrine (see Section 9.1.1), which affords the optically active compound (−)-(*S*)-2-ethyl-2-diethoxymethyl-4-pentenoic acid (**11/77**). Treatment of its ester **11/76** with diborane/H_2O_2 gives the chiral product **11/78** which is functionalized at the required C-19 position. The hydroxyl group is tosylated so as to provide a suitable leaving group, then the product is reacted with tryptamine (**11/70**) to give (+)-(*S*)-**11/80**. This substance is lactamized (in 74% yield!) in an interesting reaction by heating to 130°C in molten imidazole. The action of concentrated acetic acid on the lactam (+)-(*S*)-**11/81** releases the formyl group, which is then extended by an extra two carbon atoms by means of the above-mentioned Wittig-Horner reaction to give **11/68**. A Bischler-Napieralski cyclization in the presence of POCl₃ affords **11/83**. This quaternary compound is reduced with NaBH₄ to give exclusively the isomer **11/84** with the wrong configuration at C-3, but the yield of the required product **11/61** can be optimized by catalytic

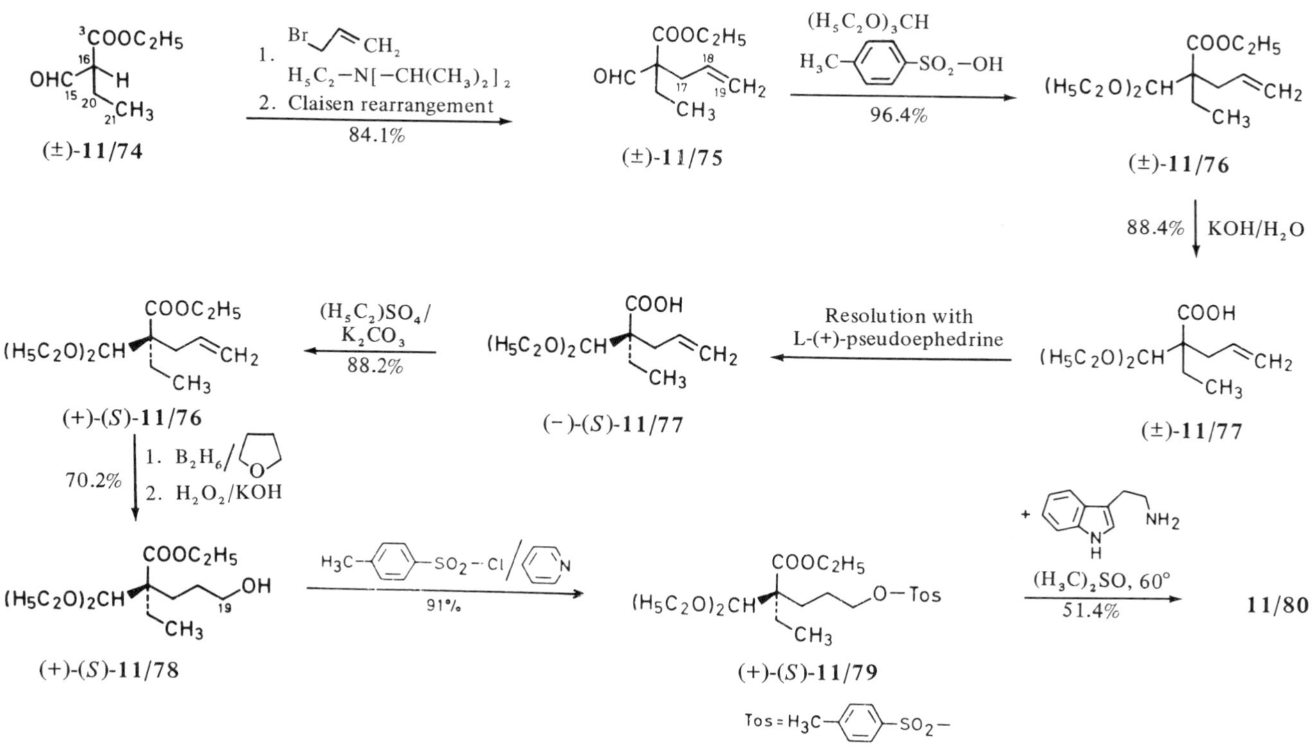

^{3}COOC$_2$H$_5$

OHC ^{16} H ^{21}CH$_3$ ^{15} ^{20}

(±)-**11/74**

1. Br~~CH$_2$

H$_5$C$_2$—N[—CH(CH$_3$)$_2$]$_2$

2. Claisen rearrangement

84.1%

COOC$_2$H$_5$

OHC ^{18} ^{17} CH$_3$ ^{19}CH$_2$

(±)-**11/75**

(H$_5$C$_2$O)$_3$CH

H$_3$C—⟨ ⟩—SO$_2$—OH

96.4%

COOC$_2$H$_5$

(H$_5$C$_2$O)$_2$CH CH$_3$ CH$_2$

(±)-**11/76**

KOH/H$_2$O

88.4%

COOH

(H$_5$C$_2$O)$_2$CH CH$_3$ CH$_2$

(±)-**11/77**

Resolution with
L-(+)-pseudoephedrine

COOH

(H$_5$C$_2$O)$_2$CH CH$_3$ CH$_2$

(−)-(S)-**11/77**

COOC$_2$H$_5$

(H$_5$C$_2$O)$_2$CH CH$_3$ CH$_2$

(+)-(S)-**11/76**

1. B$_2$H$_6$/⟨O⟩
2. H$_2$O$_2$/KOH

70.2%

COOC$_2$H$_5$

(H$_5$C$_2$O)$_2$CH CH$_3$ ^{19} OH

(+)-(S)-**11/78**

H$_3$C—⟨ ⟩—SO$_2$—Cl / ⟨N⟩

91%

COOC$_2$H$_5$

(H$_5$C$_2$O)$_2$CH CH$_3$ O—Tos

(+)-(S)-**11/79**

Tos = H$_3$C—⟨ ⟩—SO$_2$—

⟨indole⟩ NH$_2$ +

(H$_3$C)$_2$SO, 60°

51.4%

11/80

Scheme 28 Synthesis of vincamine, **11/58** [P 19].[3]

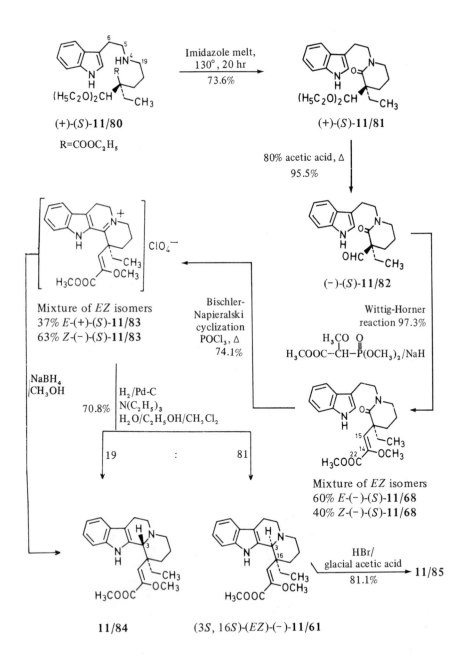

Scheme 28 Synthesis of vincamine (continued).

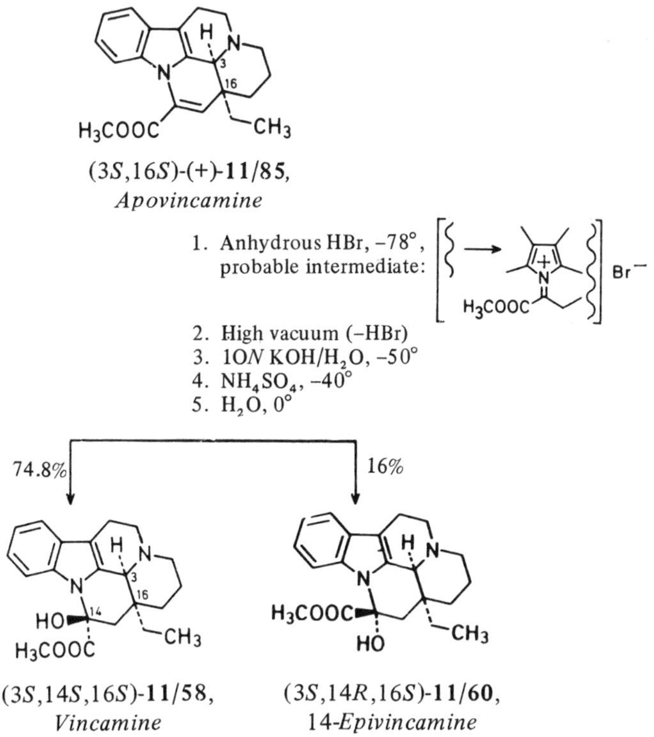
H3COOC
CH3
(3S,16S)-(+)-11/85,
Apovincamine
1. Anhydrous HBr, -78°, probable intermediate:
H3COOC
Br⁻
2. High vacuum (-HBr)
3. 10N KOH/H2O, -50°
4. NH4SO4, -40°
5. H2O, 0°
74.8%
16%
HO
H3COOC
CH3
H3COOC
HO
CH3
(3S,14S,16S)-11/58,
Vincamine
(3S,14R,16S)-11/60,
14-Epivincamine

Scheme 28 Synthesis of vincamine (concluded).

hydrogenation under special reaction conditions using a solvent mixture containing $N(C_2H_5)_3$, so that a ratio of 81 parts of **(11/61)** to 19 parts of **(11/84)** is obtained. It may be noted that the compounds **11/68**, **11/83**, and **11/61** were obtained as mixtures of *EZ* isomers, which were separated for characterization purposes but used as such in further reactions, since the difference between the isomers disappears in the final product. The protected carbonyl group in **11/61** must be liberated before cyclization to vincamine **(11/58)**, but this is only possible under drastic reaction conditions. The main product when HBr/glacial acetic acid is used is the dehydrated analogue apovincamine((+)-**11/85**) which is obtained in 81% yield while vincamine **(11/58)** and its 14-epimer **11/60** are obtained only as by-products.

Reaction conditions were finally found for the conversion of apovincamine **(11/85)** into vincamine **(11/58)**, no doubt specially devised to meet the challenge involved here. These conditions are quite special and not necessarily applicable in more general cases. Apovincamine is dissolved in absolute hydrogen bromide at −78°C, the solution is evaporated to dryness at the same temperature, then 10 *N* aqueous potassium hydroxide is added at −50°C. Under these conditions aqueous sodium hydroxide is solid. The solution is allowed to warm up slowly, with addition of ammonium sulfate around $-40°C$ and with water at $0°C$ added around $-5°C$. Vincamine **(11/58)** and its epimer **11/60** were finally obtained in a combined yield of 91% of theory, of which vincamine with 82% forms the main product (see Scheme 28).

From this example of a stereoselective synthesis it is clear that from economic and financial considerations it is advantageous to carry out the resolution as early as possible in the synthetic plant, so as to avoid having to work initially with a large quantity of substance which is halved at the end of the sequence in the process of resolution.

As mentioned previously, a number of other syntheses of vincamine **(11/58)** have been published. The reactions involved are likewise of considerable interest, but a detailed description and a comparison between them is not possible for lack of space.

Abbreviations for Plant Families

ACA	Acanthaceae		ELA	Elaeagnaceae
AIZ	Aizoaceae		ELE	Elaeocarpaceae
ALA	Alangiaceae		EPH	Ephedraceae
AMA	Amaryllidaceae		ERY	Erythroxylaceae
ANC	Ancistrocladaceae		EUP	Euphorbiaceae
ANN	Annonaceae			
APO	Apocynaceae		FLI	Flindersiaceae
ARA	Araceae		FUM	Fumariaceae
ARI	Aristolochiaceae			
ASC	Asclepiadaceae		GAR	Garryaceae
ATH	Atherospermataceae		GEN	Gentianaceae
			GYR	Gryostemonaceae
BER	Berberidaceae			
BOR	Boraginaceae		HDR	Hydrangeaceae
BUX	Buxaceae		HYD	Hydrasidaceae
			HYP	Hypecoaceae
CAC	Cactaceae			
CAL	Calycanthaceae		ICA	Icacinaceae
CAM	Campanulaceae			
CAR	Caricaceae		LAB	Labiatae
CEL	Celastraceae		LAU	Lauraceae
CEP	Cephalotaxaceae		LEG	Leguminosae
CHE	Chenopodiaceae		LEO	Leonticaceae
CLA	Clavicipidaceae		LIL	Liliaceae
COM	Compositae		LYC	Lycopodiaceae
CYP	Cyperaceae		LYT	Lythraceae

MAG	Magnoliaceae	PTE	Pteridophyllaceae
MEN	Menispermaceae	PUN	Punicaceae
MON	Monimiaceae		
MOR	Moraceae	RAN	Ranunculaceae
		RHA	Rhamnaceae
NAN	Nandinaceae	RUB	Rubiaceae
NEL	Nelumbonaceae	RUT	Rutaceae
NYM	Nymphaeaceae		
NYS	Nyssaceae	SAL	Salvadoraceae
		SCR	Scrophulariaceae
OLA	Olacaceae	SIM	Simaroubaceae
OLE	Oleaceae	SOL	Solanaceae
ORC	Orchidaceae	STR	Strychnaceae
		SYM	Symplocaceae
PAP	Papaveraceae		
PAS	Passifloraceae	UMB	Umbelliferae
PER	Periplocaceae		
PIP	Piperaceae	ZYG	Zygophyllaceae
POL	Polygonaceae		

Bibliography

A 1   Authors of collected articles on pharmacology in M 1: **5** (1955).

A 2   W. A. Ayer and T. E. Habgood, "The pyridine alkaloids" in M 1: **11**, 459 (1968).

A 3   W. A. Ayer, "The *Lycopodium* alkaloids including synthesis and biosynthesis" in H 17: p. 1.

A 4   V. Agwada, M. B. Patel, M. Hesse, and H. Schmid, "Die Alkaloide aus *Hedranthera barteri* (Hook. f.) Pichon," *Helv. Chim. Acta* **53**, 1567 (1970).

B 1   H. G. Boit, *Ergebnisse der Alkaloid-Chemie bis 1960*, Akademie-Verlag, Berlin, 1961.

B 2   A. Brossi and B. Pecherer, "Alkaloids containing a simple aromatic moiety" in P 1: p. 11.

B 3   A. Burger, "The benzylisoquinoline alkaloids" in M 1: **4**, 29 (1954).

B 4   V. Boekelheide, "The *Erythrina* alkaloids" in M 1: **7**, 201 (1960).

B 5   F. Bohlmann and D. Schumann, "Lupine alkaloids" in M 1: **9**, 176 (1967).

B 6   K. W. Bentley, "The morphine alkaloids" in P 1: p. 117.

B 7   H. Budzikiewicz, C. Djerassi, and D. H. Williams, "Structure elucidation of natural products by mass spectrometry," *Alkaloids*, Vol. 1, Holden-Day, San Francisco, 1964.

B 8   K. Biemann, "Mass spectrometry of selected natural products," *Fortschr. Chem. Org. Naturst.* **24**, 1 (1966).

B 9   K. S. Brown, "Abnormal steroidal alkaloids" in P 1: p. 631.

B 10 A. R. Battersby and H. F. Hodson, "Alkaloids of calabash curare and *Strychnos* species" in M 1: **8**, 515 (1965).

B 11 A. R. Battersby and H. F. Hodson, "Alkaloids of calabash curare and *Strychnos* species" in M 1: **11**, 189 (1968).

B 12 K. Bernauer, "Alkaloide aus Calebassencurare und südamerikanischen Strychnosarten," *Fortschr. Chem. org. Naturst.* **17**, 183 (1959).

B 13 K. Bernauer and W. Hofheinz, "Proaporphin-Alkaloide," *Fortschr. Chem. org. Naturst.* **26**, 245 (1968).

B 14 K. W. Bentley, *The Chemistry of the Morphine Alkaloids*, Oxford University Press, London, 1954.

B 15 K. W. Bentley, *The Alkaloids*, Interscience, New York, 1957.

B 16 A. R. Battersby, "Alkaloid biogenesis," *Quart. Rev.* **15**, 259 (1961).

204   Bibliography

B 17   A. R. Battersby, "Recent researches on indole alkaloids," *Pure Appl. Chem.* **6**, 471 (1963).

B 18   A. R. Battersby and H. F. Hodson, "Alkaloids of calabash-curare and *Strychnos* species," *Quart. Rev.* **14**, 77 (1960).

B 19   A. R. Battersby and H. T. Openshaw, "The imidazole alkaloids" in M 1: **3**, 201 (1953).

B 20   K. Bernauer and F. Schneider, "Alkaloide" in *Ullmanns Encyklopädie der technischen Chemie*, 4th ed., Vol. 7, Verlag Chemie, Weinheim, 1973.

B 21   H.-G. Boit, "*chano*-Basen der *N*-Methyl-pseudobrucin-Reihe" (IV. Mitteil. über Strychnos-Alkaloide), *Chem. Ber.* **84**, 923 (1951).

B 22   H. Budzikiewicz, "Zur Problematik der massenspektrometrischen Untersuchung organischer Verbindungen: Umwandlungen vor der Ionisierung," *Z. Anal. Chem.* **244**, 1 (1969).

B 23   M. M. Badawi, K. Bernauer, P. van den Broek, D. Gröger, A. Guggisberg, S. Johne, I. Kompiš, F. Schneider, H.-J. Veith, M. Hesse, and H. Schmid, "Macrocyclic spermidine and spermine alkaloids," *Pure Appl. Chem.* **33**, 81 (1973).

B 24   A. R. Battersby, T. H. Brown, and J. H. Clements, "Syntheses along biosynthetic pathways. Part I. Synthesis of (+)-isothebaine," *J. Chem. Soc.* 4550 (1965).

B 25   D. H. R. Barton, G. W. Kirby, W. Steglich, and G. W. Thomas, "The biosynthesis and synthesis of morphine alkaloids," *Proc. Chem. Soc.* 203 (1963).

B 26   A. R. Battersby, R. Southgate, J. Staunton, and M. Hirst, "Synthesis of (+)- and (–)-scoulerine and of (+)- and (–)-coreximine," *J. Chem. Soc. C* 1052 (1966).

B 27   A. R. Battersby and B. J. T. Harper, "Biogenesis of morphine," *Chem. & Ind.* 364 (1958).

B 28   A. R. Battersby, R. Binks, and D. J. Le Count, "Biosynthesis of morphine," *Proc. Chem. Soc.* 287 (1969).

B 29   K. Bláha, Z. Koblicová, and J. Trojánek, "Chiroptical investigation on the *Iboga* and *Voacanga* alkaloids," *Coll. Czech. Chem. Commun.* **39**, 2258 (1974).

B 30   F. von Bruchhausen and H. W. Bersch, "Über die Konstitution des Chelidonins," *Ber. dtsch. chem. Ges.* **63**, 2520 (1930).

B 31   H. Bickel, H. Schmid, and P. Karrer, "Zur Kenntnis des Fluorocurins und Mavacurins," *Helv. Chim. Acta* **38**, 649 (1955).

B 32   K. Bodendorf and W. Krieger, "Über die Alkaloide von *Mesembryanthemum tortuosum* L.," *Arch. Pharm.* **290**, 441 (1957).

B 33   K. W. Bentley, "The morphine alkaloids" in M 1: **13**, 3 (1971).

B 34   A. Brossi, S. Teitel, and G. V. Parry, "The Ipecac alkaloids" in M 1: **13**, 189 (1971).

B 35   J. S. Bindra, "Oxindole alkaloids" in M 1: **14**, 84 (1973).

C 1   J. W. Cook and J. D. Loudon, "Alkaloids of the Amaryllidaceae" in M 1: **2**, 331 (1952).

C 2   M. Curcumelli-Rodostamo and M. Kulka, "Bisbenzylisoquinoline and related alkaloids" in M 1: **9**, 133 (1967).

C 3   A. Chatterjee and G. Ganguli, "The chemistry of bisindole alkaloids," *J. Sci. Ind. Res. (India)* **23**, 178 (1964).

C 4   E. Coxworth, "Alkaloids of the calabar bean" in M 1: **8**, 27 (1965).

C 5   J. W. Cook and J. D. Loudon, "Colchicine" in M 1: **2**, 261 (1952).

C 6   B. T. Cromwell, "The alkaloids" in K. Paech and M. V. Tracey; Eds., *Moderne Methoden der Pflanzenanalyse*, Vol. 4, Springer-Verlag, Berlin, 1955, p. 367.

C 7   E. G. C. Clarke, "The forensic chemistry of alkaloids" in M 1: **12**, 514 (1970).

C 8   V. Černy and F. Šorm, "Steroid alkaloids: Alkaloids of Apocynaceae and Buxaceae" in M 1: **9**, 305 (1967).

C 9   A. Chatterjee, "*Rauwolfia* alkaloids," *Fortschr. Chem. org. Naturst.* **10**, 390 (1953).

C 10   A. Chatterjee and S. C. Pakrashi, "Recent developments in the chemistry and pharmacology of *Rauwolfia* alkaloids," *Fortschr. Chem. org. Naturst.* **13**, 346 (1956).

C 11   "The alkaloids," (annual review), Specialist Periodical Reports, The Chemical Society, London, Vol. 1 (Literature 1969-1970) 1971; Vol. 2 (1970-1971) 1972; Vol. 3 (1971-1972) 1973; Vol. 4 (1972-1973), 1974.

C 12   S. S. Cohen, *Introduction to the polyamines*, Prentice-Hall, Englewood Cliffs, NJ, 1971.

C 13   M. P. Cava, S. S. Tjoa, Q. A. Ahmed, and A. I. Da Rocha, "The alkaloids of *Tabernaemontana riedelii* and *T. rigida*," *J. Org. Chem.* **33**, 1055 (1968).

C 14   A. C. Cope and A. A. D'Addieco, "Cyclic polyolefins. XVI. Phenylcycloheptatriene and phenylcyclooctatriene," *J. Am. Chem. Soc.* **73**, 3419 (1951).

C 15   T. J. Curphey and H. L. Kim, "A new synthetic approach to the Amaryllidaceae alkaloids. Application to the synthesis of mesembrine and mesembrinine," *Tetrahedron Lett.* 1441 (1968).

C 16   E. J. Corey and R. D. Balanson, "A total synthesis of (±)-porantherine," *J. Am. Chem. Soc.* **96**, 6516 (1974).

C 17   M. Curcumelli-Rodostamo, "Bisbenzylisoquinoline and related alkaloids" in M 1: **13**, 304 (1971).

D 1   V. Deulofeu, J. Comín, and M. J. Vernengo, "The benzylisoquinoline alkaloids" in M 1: **10**, 402 (1968).

D 2   G. Dalma, "The *Erythrophleum* alkaloids" in M 1: **4**, 265 (1954).

D 3   C. Djerassi, H. Budzikiewicz, J. M. Wilson, J. Gosset, J. Le Men, and M.-M. Janot, "Mass spectrometry in structural and stereochemical problems. Vincadifformine (Alcaloïdes des pervenches)," *Tetrahedron Lett.* 235 (1962).

D 4   R. W. Doskotch, A. B. Ray, W. Kubelka, E. H. Fairchild, C. D. Huffort, and J. L. Beal, "The structure of codonocarpine," *Tetrahedron* **30**, 3229 (1974).

D 5   W. A. Denne, S. R. Johns, J. A. Lamberton, and A. McL. Mathieson, "The absolute structure of porantherine," *Tetrahedron Lett.* 3107 (1971).

D 6   W. A. Denne and A. McL. Mathieson, "*Poranthera corymbosa* alkaloids. I. Crystal and absolute molecular structure of porantherine hydrobromide," *J. Cryst. Mol. Struct.* **3**, 79 (1973).

D 7   W. A. Denne, S. R. Johns, J. A. Lamberton, A. McL. Mathieson, and H. Suares, "The molecular structure and absolute configuration of poranthericine and porantheridine," *Tetrahedron Lett.* 1767 (1972).

D 8   W. A. Denne and A. McL. Mathieson, "*Poranthera corymbosa* alkaloids. II. Molecular structure and absolute configuration of porantheridine hydrobromide," *J. Cryst. Mol. Struct.* **3**, 87 (1973).

D 9   W. A. Denne and A. McL. Mathieson, "*Poranthera corymbosa* alkaloids. III. Molecular structure and absolute configuration of poranthericine hydrobromide," *J. Cryst. Mol. Struct.* **3**, 139 (1973).

D 10   W. A. Denne, "*Poranthera corymbosa* alkaloids. IV. Crystal and absolute structure of 4-methyl-6-(2'-benzoyloxypentyl) quinolizidine," *J. Cryst. Mol. Struct.* 3, 367 (1973).

E 1   J. Ewing, G. K. Hughes, E. Ritchie, and W. C. Taylor, "The alkaloids of *Cryptocarya bowiei* (Hook.) Druce," *Aust. J. Chem.* 6, 78 (1953).

F 1   G. Fodor, "Tropane alkaloids" in P 1: p. 431.

F 2   G. Fodor, "The tropane alkaloids" in M 1: 6, 145 (1960).

F 3   G. Fodor, "The tropane alkaloids" in M 1: 9, 269 (1967).

F 4   H.-G. Floss, U. Mothes, and A. Rittig, "Die Beziehung zwischen Gentianin und Gentiopikrosid," *Z. Naturforsch.* 19 b, 1106 (1964).

F 5   E. Fujita and T. Tomimatsu, "Alkaloids of *Thalictrum thunbergii* IV. Structure of thalicthuberine, a tertiary base in the root," *Yakugaku Zasshi* 79, 1252 (1959); *Chem Abstr.* 54, 4643.

F 6   G. Fodor, "The tropane alkaloids" in M 1: 13, 352 (1971).

F 7   C. Fuganti, "The Amaryllidaceae alkaloids" in M 1: 15, 83 (1975).

F 8   G. Fodor, "New methods and recent developments of the stereochemistry of ephedrine, pyrrolizidine, granatane and tropane alkaloids," in G. Fodor, Ed., *Recent Developments in the Chemistry of Natural Carbon Compounds*, Vol. 1, Hungarian Academy of Sciences, Budapest, 15 (1965).

G 1   B. Gilbert, "The alkaloids of *Aspidosperma, Diplorrhyncus, Kopsia, Ochrosia, Pleiocarpa, Melodinus* and related genera" in M 1: 8, 336 (1965).

G 2   B. Gilbert, "The alkaloids of *Aspidosperma, Diplorrhyncus, Kopsia, Ochrosia, Pleiocarpa, Melodinus* and related genera" in M 1: 11, 205 (1968).

G 3   D. Ginsberg, *The Opium alkaloids*, Interscience, New York, 1962.

G 4   T. R. Govindachari, "*Tylophora* alkaloids" in M 1: 9, 517 (1967).

G 5   D. Gröger, "Fortschritte der Chemie und Biochemie der Mutterkornalkaloide," *Fortschr. chem. Forsch.* 6, 159 (1966).

G 6   F. Galinovsky, "Lupinen-Alkaloide und verwandte Verbindungen," *Fortschr. Chem. org. Naturst.* 8, 245 (1951).

G 7   M. Guggenheim, *Die biogenen Amine*, Basel-Verlag von S. Karger, New York, 1940.

G 8   A. A. Gorman, M. Hesse, and H. Schmid, "Bisindole alkaloids," *The alkaloids*, Vol. 1, Specialist Periodical Reports, The Chemical Society, London, 1971, p. 201.

G 9   D. Gross, "Naturstoffe mit Pyridinstruktur und ihre Biosynthese," *Fortschr. Chem. org. Naturst.* 28, 109 (1970).

G 10   T. R. Govindachari, S. S. Sathe, and N. Viswanathan, "Gentianine, an artefact in *Enicostemma littorale*," *Indian J. Chem.* 4, 201 (1966).

G 11   D. Ganzinger and M. Hesse, "A chemotaxonomic study of the subfamily Plumerioideae of the Apocynaceae," *Lloydia* 39, 326 (1976).

G 12   M. Gorman, N. Neuss, and N. J. Cone, "*Vinca* alkaloids. XVII. Chemistry of catharanthine," *J. Am. Chem. Soc.* 87, 93 (1965).

G 13   T. R. Govindachari, B. R. Pai, and K. Nagarajan, "Chemical examination of *Tylophora asthmatica*. Part I," *J. Chem. Soc.* 2801 (1954).

G 14   T. R. Govindachari, M. V. Lakshmikantham, K. Nagarajan, and B. R. Pai, "Chemical examination of *Tylophora asthmatica*-II," *Tetrahedron* 4, 311 (1958).

G 15   J. C. Godfrey, D. S. Tarbell, and V. Boekelheide, "The structure of α-erythroidine," *J. Am. Chem. Soc.* 77, 3342 (1955).

G 16   T. R. Govindachari, N. S. Narsimhan, and S. Rajadurai, "Some degradation studies of carpaine," *J. Chem. Soc.* 558, (1957).

G 17   A. Guggisberg, M. M. Badawi, M. Hesse, and H. Schmid, "Über die Struktur der makrocyclischen Spermidin-Alkaloide Oncinotin, Neooncinotin und Isooncinotin," *Helv. Chim. Acta* 57, 414 (1974).

G 18   A. Guggisberg, P. van den Broek, M. Hesse, H. Schmid, F. Schneider, and K. Bernauer, "Synthese der makrocyclischen Spermidin-Alkaloide Oncinotin, Neooncinotin, Isooncinotin and Pseudooncinotin in racemischer Form," *Helv. Chim. Acta* 59, 3013 (1976).

G 19   D. Gross, "Vorkommen, Struktur und Biosynthese natürlicher Piperidinverbindungen," *Fortschr. Chem. Org. Naturst.* 29, 1 (1971).

G 20   K. H. Gibson and J. E. Saxton, "Total synthesis of (±)-vincamine," *Chem. Commun.* 1490 (1969).

H 1   T. A. Henry, *The Plant Alkaloids*, 4th ed., Churchill, London, 1949.

H 2   R. Hegnauer, *Chemotaxonomie der Pflanzen. Eine Übersicht über die Verbrietung und die systematische Bedeutung der Pflanzenstoffe*, Birkhäuser-Verlag, Basel, Vol. 1, 1962; Vol. 2, 1963; Vol. 3, 1964; Vol. 4, 1966; Vol. 5, 1969.

H 3   J. Holubek and O. Strouf, *Spectral Data and Physical Constants* of Alkaloids, Heyden, London, 1965.

H 4   R. K. Hill, "The *Erythrina* alkaloids" in M 1: 9, 483 (1967).

H 5   R. K. Hill, "Pyrrolidine, piperidine, pyridine and imidazole alkaloids" in P 1: p. 385.

H 6   H. L. Holmes, "The chemistry of the tropane alkaloids" in M 1: 1, 271 (1950).

H 7   H. L. Holmes, "The morphine alkaloids I" in M 1: 2, 1 (1952).

H 8   H. L. Holmes and G. Stork, "The morphine alkaloids II" in M 1: 2, 161 (1952).

H 9   H. L. Holmes, "Sinomenine" in M 1: 2, 219 (1952).

H 10   G. Habermehl, "The steroid alkaloids: the *Salamandra* group" in M 1: 9, 427 (1967).

H 11   H. L. Holmes, "The *Strychnos* alkaloids" in M 1: 1, 375 (1950).

H 12   H. L. Holmes, "The Strychnos alkaloids" in M 1: 2, 513 (1952).

H 13   J. B. Hendrickson, "The *Strychnos* alkaloids" in M 1: 6, 179 (1960).

H 14   M. Hesse, *Indolakaloide in Tabellen*, Springer-Verlag, Berlin-Heidelberg-New York, 1964.

H 15   M. Hesse, *Indolalkaloide in Tabellen*, Springer-Verlag, Berlin-Heidelberg-New York, supplement, 1968.

H 16   M. Hesse, *Progress in Mass Spectrometry*, Vol. 1, *Indolalkaloide*, Verlag Chemie, Weinheim, 1974.

H 17   D. H. Hey, Ed., *MTP International Review of Science*, Organic Chemistry Ser. I, Vol. 9, *Alkaloids*, Butterworth-University Park Press, London-Baltimore, 1973.

H 18   G. Habermehl, "Steroid alkaloids" in H 17: p. 235.

H 19   N. K. Hart, S. R. Johns, and J. A. Lamberton, "Tertiary alkaloids of *Alstonia spectabilis* and *Alstonia glabriflora* (Apocynaceae)," *Aust. J. Chem.* 25, 2739 (1972).

H 20   M. Hesse and H. Schmid, "Macrocyclic spermidine and spermine alkaloids," *MTP International Review of Science*, Organic Chemistry Ser. II, Vol. 9, *Alkaloids*, Butterworths, London, 1976, p. 265.

H 21   M. Hesse, H. Hürzeler, C. W. Gemenden, B. S. Joshi, W. I. Taylor, and H. Schmid, "Die Struktur des *Alstonia*-Alkaloides Villalstonin," *Helv. Chim. Acta* 48, 689 (1965).

H 22   A. W. Hofmann "Einwirkung der Wärme auf die Ammoniumbasen," *Ber. dtsch. chem. Ges.* **14**, 659 (1881).

H 23   R. D. Haworth, J. McKenna, and R. G. Powell, "The constitution of conessine. Part V. Synthesis of some basic steroids," *J. Chem. Soc.* 1110 (1953).

H 24   M. Hesse and H. O. Bernhard, *Progress in mass spectrometry*, Vol. 3, *Alkaloide*, Verlag Chemie, Weinheim, 1975.

H 25   J. L. Herrmann, R. J. Cregge, J. E. Richman, C. L. Semmelhack, and R. H. Schlessinger, "A high yield stereospecific total synthesis of vincamine," *J. Am. Chem. Soc.* **96**, 3702 (1974).

H 26   G. Hugel, J. Lévy, and J. Le Men, "Méthylène-indolines, indolénines et indoléniniums VI. Action de réactifs oxydants. Hemisynthèse de la vincamine," *C. R. Acad. Sci.* **274**, 1350 (1972).

I 1   R. E. Ireland, *Organic synthesis*, Prentice-Hall, Englewood Cliffs, NJ, 1969.

I 2   S. Itô, "Alkaloids" in K. Nakanishi, *Natural Products Chemistry*, Vol. 2, Kodansha Ltd., Tokyo Academic Press, New York, 1975, p. 255.

J 1   M.-M. Janot, "The Ipecac alkaloids" in M 1: **3**, 363 (1953).

J 2   P. W. Jeffs, "The protoberberine alkaloids" in M 1: **9**, 41 (1967).

J 3   O. Jeger and V. Prelog, "Steroid alkaloids: The *Holarrhena* group" in M 1: **7**, 319 (1960).

J 4   O. Jeger and V. Prelog, "Steroid alkaloids: *Veratrum* group" in M 1: **7**, 363 (1960).

J 5   P. W. Jeffs, "The Amaryllidaceae alkaloids" in H 17: p. 273.

J 6   S. Johne, D. Gröger, and M. Hesse, "Neue Alkaloide aus *Adhatoda vasica* Nees," *Helv. Chim. Acta* **54**, 826 (1971).

J 7   S. R. Johns. J. A. Lamberton, A. A. Sioumis, and H. Suares, "The alkaloids of *Poranthera corymbosa* (Euphorbiaceae)," *Aust. J. Chem.* **27**, 2025 (1974).

J 8   S. R. Johns and J. A. Lamberton, "*Elaeocarpus* alkaloids" in M 1: **14**, 326 (1973).

K 1   M. Kulka, "Bisbenzylisoquinoline alkaloids" in M 1: **7**, 439 (1960).

K 2   M. Kulka, "Bisbenzylisoquinoline alkaloids" in M 1: **4**, 199 (1954).

K 3   L. H. Keith and S. W. Pelletier, "The C_{19}-diterpene alkaloids" in P 1: p. 549.

K 4   S. M. Kupchan and A. W. By, "The steroid alkaloids: The *Veratrum* group" in M 1: **10**, 193 (1968).

K 5   I. Kompiš, M. Hesse, and H. Schmid, "An approach to the biogenetic classification of indole alkaloids," *Lloydia* **34**, 269 (1971).

K 6   T. Kametani, *The Chemistry of the Isoquinoline Alkaloids*, Hirokawa Publishing-Elsevier, Tokyo-Amsterdam, 1969.

K 7   J. P. Kutney, "The total synthesis of indole alkaloids" in H 17: p. 27.

K 8   T. Kametani and K. Fukumoto, "Benzylisoquinoline and homobenzylisoquinoline alkaloids" in H 17: p. 181.

K 9   J. P. Kutney, R. T. Brown, and E. Piers, "The absolute configuration of the *Iboga* alkaloids," *Can. J. Chem.* **44**, 637 (1966).

K 10   T. Kishi, M. Hesse, W. Vetter, C. W. Gemenden, W. I. Taylor, and H. Schmid, "Macralstonin," *Helv. Chim. Acta* **49**, 946 (1966).

K 11   J. R. Know and J. Slobbe, "Three novel alkaloids from *Ervatamia orientalis*," *Tetrahedron Lett.* 2149 (1971).

K 12   S. L. Keely and F. C. Tahk, "The 3-arylpyrrolidine alkaloid synthon. A new synthesis of *dl*-mesembrine," *J. Am. Chem. Soc.* **90**, 5584 (1968).

K 13   R. S. Kapil, "The Carbazole Alkaloids" in M 1: **13**, 273 (1971).

K 14   T. Kametani and M. Koizumi, "Phenethyl isoquinoline alkaloids" in M 1: **14**, 265 (1973).

K 15   M. E. Kuehne, "The total synthesis of vincamine," *Lloydia* **27**, 435 (1964); *J. Am. Chem. Soc.* **86**, 2946 (1964).

L 1   E. Leete, "Alkaloid biogenesis" in P. Bernfeld, *Biogenesis of Natural Compounds*, 2nd ed., Pergamon, Oxford, 1966, p. 953.

L 2   H. W. Liebisch, "Cyclisierungsmechanismen bei der Alkaloid-Biosynthese," *Fortschr. Chem. Forsch.* **9**, 534 (1968).

L 3   N. J. Leonard, "*Lupin* alkaloids" in M 1: **3**, 119 (1953).

L 4   N. J. Leonard, "*Lupin* alkaloids" in M 1: **7**, 253 (1960).

L 5   N. J. Leonard, "*Senecio* alkaloids" in M 1: **1**, 107 (1950).

L 6   N. J. Leonard, "*Senecio* alkaloids" in M 1: **6**, 35 (1960).

L 7   E. Leete, "Alkaloid biosynthesis," *Ann. Rev. Plant Physiol.* **18**, 179 (1967).

L 8   B. Lythgoe, "The *Taxus* alkaloids" in M 1: **10**, 597 (1968).

M 1   R. H. F. Manske, *The Alkaloids—Chemistry and Physiology*, Vols. 1-15, Academic, New York, 1950-1975.

M 2   R. H. F. Manske, "The aporphine alkaloids" in M 1: **4**, 119 (1954).

M 3   R. H. F. Manske and W. A. Harrison, "The alkaloids of *Geissospermum* species" in M 1: **8**, 679 (1965).

M 4   K. Mothes and H. R. Schütte, *Biosynthese der Alkaloide,* VEB Deutscher Verlag der Wissenschaften, Berlin, 1969.

M 5   R. H. F. Manske, "The alkaloids of Calycanthaceae" in M 1: **8**, 581 (1965).

M 6   R. H. F. Manske, "The carboline alkaloids" in M 1: **8**, 47 (1965).

M 7   R. H. F. Manske, "The quinazolinocarbolines" in M 1: **8**, 55 (1965).

M 8   J. McKenna, "Steroidal alkaloids," *Quart. Rev.* **7**, 231 (1953).

M 9   A. Mondon, "*Erythrina* alkaloids" in P 1: p. 173.

M 10   L. Marion, "The *Erythrina* alkaloids" in M 1: **2**, 499 (1952).

M 11   R. B. Morin, "*Erythrophleum* alkaloids" in M 1: **10**, 287 (1968).

M 12   R. H. F. Manske, "*Isoquinoline* alkaloids" in M 1: **7**, 423 (1960).

M 13   R. H. F. Manske, "The biosynthesis of isoquinolines" in M 1: **4**, 1 (1954).

M 14   L. Marion, "The indole alkaloids" in M 1: **2**, 369 (1952).

M 15   R. H. F. Manske, "The Ipecac alkaloids" in M 1: **7**, 419 (1960).

M 16   D. B. MacLean, "The *Lycopodium* alkaloids" in P 1: p. 469.

M 17   D. B. MacLean, "The Lycopodium alkaloids" in M 1: **10**, 306 (1968).

M 18   R. H. F. Manske, "The *Lycopodium* alkaloids" in M 1: **7**, 505 (1960).

M 19   R. H. F. Manske, "α-Naphthaphenanthridine alkaloids" in M 1: **4**, 253 (1954).

M 20   R. H. F. Manske, "α-Naphthaphenanthridine alkaloids" in M 1: **10**, 485 (1968).

M 21   R. H. F. Manske, "The protopine alkaloids" in M 1: **4**, 147 (1954).

M 22   R. H. F. Manske and W. R. Ashford, "The protoberberine alkaloids" in M 1: **4**, 77 (1954).

M 23   L. Marion, "The pyrrolidine alkaloids" in M 1: **1**, 91 (1950).

M 24   L. Marion, "The pyrrolidine alkaloids" in M 1: **6**, 31 (1960).

M 25   L. Marion, "The pyridine alkaloids" in M 1: **1**, 165 (1950).

M 26 L. Marion, "The pyridine alkaloids" in M 1: **6**, 123 (1960).

M 27 R. H. F. Manske, "Papaveraceae alkaloids" in M 1: **10**, 467 (1968).

M 28 R. H. F. Manske, "Alkaloids of *Pseudocinchona* and *Yohimbe*" in M 1: 8, 694 (1965).

M 29 H. J. Monteiro, "Yohimbine and related alkaloids" in M 1: **11**, 145 (1968).

M 30 K. J. Morgan and J. A. Barltrop, "*Veratrum* alkaloids," *Quart. Rev.* **12**, 34 (1958).

M 31 R. H. F. Manske, "The cularine alkaloids" in M 1: **4**, 249 (1954).

M 32 R. H. F. Manske, "The cularine alkaloids" in M 1: **10**, 463 (1968).

M 33 S. McLean and J. Whelan, "Spirobenzylisoquinoline alkaloids" in H 17: p. 161.

M 34 R. J. Miller, C. Jolles, and H. Rapoport, "Morphine metabolism and normorphine in *Papaver somniferum*," *Phytochem.* **12**, 597 (1973).

M 35 H. Mehri, M. Plat, and P. Potier, "Plantes de Nouvelle-Calédonie V. *Mélodinus scandens* Forst. Isolement de dix alcaloïdes monomères. Description de deux alcaloïdes nouveaux: *N*-oxy-épiméloscine et méloscandonine," *Ann. Pharm. Fr.* **29**, 291 (1971).

M 36 J. Mokrý, I. Kompiš, L. Dúbravková, and P. Šefčovič, "Vincadifformin und Minovincin, zwei weitere razemische Alkaloide aus *Vinca minor* L." *Experientia* **19**, 311 (1963).

M 37 N. J. McCorkindale, D. S. Magrill, M. Martin-Smith, S. J. Smith, and J. B. Stenlake, "Petaline: A 7,8-dioxygenated benzylisoquinoline," *Tetrahedron Lett.* 3841 (1964).

M 38 A. W. McKenzie and J. R. Price, "The alkaloids of *Gyrocarpus americanus* Jacq.," *Aust. J. Chem.* **6**, 180 (1953).

M 39 E. Mosettig and E. Meitzner, "An improved method for the preparation of morphenol (3-hydroxy-4,5-phenanthrylene oxide) from morphine," *J. Am. Chem. Soc.* **56**, 2738 (1934).

M 40 D. B. MacLean, R. H. F. Manske, and L. Marion, "Alkaloids of *Lycopodium* species. XI. Nature of the oxygen atom in lycopodine; some reactions of the base," *Can. J. Res.* **28B**, 460 (1950).

M 41 D. B. MacLean, "*The Lycopodium alkaloids*" in M 1: **14**, 348 (1973).

M 42 N. M. Mollov, H. B. Dutschewska, and V. S. Georgiev, "*Thalictrum* alkaloids," *Recent Developments in the Chemistry of Natural Carbon Compounds* **4**, 193 (1971).

N 1 N. Neuss, "Indole alkaloids" in P 1: p. 213.

N 2 C. R. Narayanan, "Newer developments in the field of *Veratrum* alkaloids," *Fortschr. Chem. org. Naturst.* **20**, 298 (1962).

N 3 C. E. Nordman and S. K. Kumra, "The structure of villalstonine," *J. Am. Chem. Soc.* **87**, 2059 (1965).

O 1 H. T. Openshaw, "The Ipecacuanha alkaloids" in P 1: p. 85.

O 2 H. T. Openshaw, "Quinoline alkaloids, other than those of *Cinchona*" in M 1: 3, 65 (1953).

O 3 H. T. Openshaw, "Quinoline alkaloids, other than those of *Cinchona*" in M 1: 7, 229 (1960).

O 4 H. T. Openshaw, "Quinoline alkaloids, other than those of *Cinchona*" in M 1: 9, 223 (1967).

O 5 H. T. Openshaw, "The quinazoline alkaloids" in M 1: 3, 101 (1953).

O 6 H. T. Openshaw, "The quinazoline alkaloids" in M 1: 7, 247 (1960).

P 1 S. W. Pelletier, *Chemistry of the Alkaloids*, Van Nostrand Reinhold, New York, 1970.

P 2   J. R. Price, "*Acridine* alkaloids" in M 1: **2**, 353 (1952).

P 3   S. W. Pelletier and L. H. Keith, "The C_{20}-diterpene alkaloids" in P 1: p. 503.

P 4   S. W. Pelletier and L. H. Keith, "Diterpene alkaloids from *Aconitum, Delphinium* and *Garrya* species" in M 1: **12**, 1 (1970).

P 5   S. W. Pelletier and L. H. Keith, "Diterpene alkaloids from *Aconitum, Delphinium* and *Garrya* species" in M 1: **12**, 136 (1970).

P 6   A. Popelak and G. Lettenbauer, "The mesembrine alkaloids" in M 1: **9**, 467 (1967).

P 7   V. Prelog and O. Jeger, "The chemistry of *Solanum* and *Veratrum* alkaloids" in M 1: **3**, 247 (1953).

P 8   V. Prelog and O. Jeger, "Steroid alkaloids. The *Solanum* group" in M 1: **7**, 343 (1960).

P 9   S. W. Pelletier, "The chemistry of the C_{20}-diterpene alkaloids," *Quart. Rev.* **21**, 525 (1967).

P 10   A. R. Pinder, "Lactonic alkaloids," *Chem. Rev.* **64**, 551 (1964).

P 11   J. R. Price, "Alkaloids related to anthranilic acid," *Fortschr. Chem. org. Naturst.* **13**, 302 (1956).

P 12   S. W. Pelletier and S. W. Page, "The structure and synthesis of C_{19}-diterpene alkaloids" in H 17: p. 319.

P 13   H. I. Parker, G. Blaschke, and H. Rapoport, "Biosynthetic conversion of thebaine to codeine," *J. Am. Chem. Soc.* **94**, 1276 (1972).

P 14   M. Plat, J. Le Men, and M.-M. Janot, "Structure de quatre alcaloïdes de la petite pervenche (*Vinca minor* L.): (–)-vincadiformine, minovincine, minovincinine et méthoxy-16-minovincine," *Bull. Soc. Chim. Fr.* 2237 (1962).

P 15   W. H. Perkin "Cryptopine and protopine," *J. Chem. Soc.* **109**, 815 (1916).

P 16   V. Prelog, B. C. McKusick, J. R. Merchant, S. Julia, and M. Wilhelm, "*Erythrina*-Alkaloide. Über den Bromcyan-Abbau des Dihydro-erystrins; ein Beitrag zur Konstitutionsaufklärung der aromatischen *Erythrina*-Alkaloide," *Helv. Chim. Acta* **39**, 498 (1956).

P 17   A. Popelak, G. Lettenbauer, E. Haack, and H. Spingler, "Die Struktur des Mesembrins und Mesembrenins," *Naturwiss.* **47**, 231 (1960).

P 18   V. Preininger, "The pharmacology and toxicology of the Papaveraceae alkaloids" in M 1: **15**, 207 (1975).

P 19   P. Pfäffli, W. Oppolzer, R. Wenger, and H. Hauth, "Stereoselektive Synthese von optisch aktivem Vincamin," *Helv. Chim. Acta* **58**, 1131 (1975).

R 1   R. F. Raffauf, *A handbook of Alkaloids and Alkaloid-Containing Plants*, Wiley, New York, 1970.

R 2   L. Reti, "β-Phenethylamines" in M 1: **3**, 313 (1953).

R 3   L. Reti, "*Ephedra* bases" in M 1: **3**, 339 (1953).

R 4   B. Robinson, "Alkaloids of the calabar bean" in M 1: **10**, 383 (1968).

R 5   E. Ritchie and W. C. Taylor, "The *Galbulimia* alkaloids" in M 1: **9**; 529 (1967).

R 6   L. Reti, "Simple isoquinoline alkaloids" in M 1: **4**, 7 (1954).

R 7   L. Reti, "Cactus alkaloids" in M 1: **4**, 23 (1954).

R 8   R. F. Raffauf and M. B. Flagier, "Alkaloids of the Apocynaceae," *Econ. Bot.* **14**, 37 (1960).

R 9   A. J. B. Robertson, "Field ionisation" in A. Maccoll, Ed., *MTP International Review of Science*, Vol. 5, *Mass Spectrometry*, Butterworth-University Park Press, London-Baltimore, 1972, p. 103.

R 10   R. Rodrigo, "The cancentrine alkaloids" in M 1: **14**, 407 (1973).

R 11   B. Robinson, "Alkaloids of the calabar bean" in M 1: **13**, 213 (1971).

S 1   G. A. Swan, *An Introduction to the Alkaloids*, Blackwell Scientific Publ., Oxford, 1967.

S 2   M. Shamma, "The aporphine alkaloids" in M 1: 9, 1 (1967).

S 3   J. E. Saxton, "Alkaloids of *Alstonia* species" in M 1: **12**, 207 (1970).

S 4   J. E. Saxton, "Alkaloids of *Alstonia* species" in M 1: 8, 159 (1965).

S 5   J. E. Saxton, "The simple bases" in M 1: 8, 1 (1965).

S 6   J. E. Saxton, "Alkaloids of *Haplophyton cimicidum*" in M 1: 8, 673 (1965).

S 7   I. D. Spenser, "Biosynthesis of alkaloids" in P 1: p. 669.

S 8   W. Solomon, "The *Cinchona* alkaloids" in P 1: p. 301.

S 9   E. S. Stern, "The diterpenoid alkaloids from *Aconitum, Delphinium*, and *Garrya* species" in M 1: **7**, 473 (1960).

S 10   E. S. Stern, "The *Aconitum* and *Delphinium* alkaloids" in M 1: **4**, 275 (1954).

S 11   P. J. Scheuer, "The furoquinoline alkaloids" in P 1: p. 355.

S 12   J. E. Saxton, "Alkaloids of *Gelsemium* species" in M 1: 8, 93 (1965).

S 13   J. E. Saxton, "The indole alkaloids" in M 1: **7**, 1 (1960).

S 14   J. E. Saxton, "The simple indole bases" in M 1: **10**, 491 (1968).

S 15   M. Shamma, "The isoquinoline alkaloids" in P 1: p. 31.

S 16   J. E. Saxton, "Alkaloids of *Mitragyna* and *Ourouparia* species" in M 1: 8, 59 (1965).

S 17   J. E. Saxton, "Alkaloids of *Mitragyna* and *Ourouparia* species" in M 1: **10**, 521 (1968).

S 18   G. Stork, "The morphine alkaloids" in M 1: **6**, 219 (1960).

S 19   A. Stoll and A. Hofmann, "The chemistry of the ergot alkaloids" in P 1: p. 267.

S 20   A. Stoll and A. Hofmann, "The ergot alkaloids" in M 1: 8, 726 (1965).

S 21   J. E. Saxton, "Alkaloids of *Picralima nitida*" in M 1: 8, 119 (1965).

S 22   J. E. Saxton, "Alkaloids of *Picralima nitida*" in M 1: **10**, 501 (1968).

S 23   J. Staněk, "Phthalideisoquinoline alkaloids" in M 1: **4**, 433 (1954).

S 24   J. Staněk, "Phthalideisoquinoline alkaloids" in M 1: 9, 117 (1967).

S 25   J. Staněk, and R. H. F. Manske, "Phthalideisoquinoline alkaloids" in M 1: **4**, 167 (1954).

S 26   F. Šantavý, "Papaveraceae alkaloids" in M 1: **12**, 333 (1970).

S 27   Y. Sato, "Steroidal alkaloids" in P 1: p. 591.

S 28   K. Schreiber, "Steroid alkaloids: The *Solanum* group" in M 1: **10**, 1 (1968).

S 29   G. F. Smith, "*Strychnos* alkaloids" in M 1: 8, 592 (1965).

S 30   E. Schlittler, "*Rauwolfia* alkaloids with special reference to the chemistry of reserpine" in M 1: 8, 287 (1965).

S 31   B. B. Stowe, "Occurrence and metabolism of simple indoles in plants," *Fortschr. Chem. org. Naturst.* **17**, 248 (1959).

S 32   A Stoll, "Recent investigations on ergot alkaloids," *Fortschr. Chem. org. Naturst.* **9**, 114 (1952).

S 33 A. W. Sangster and K. L. Stuart, "Ultraviolet spectra of alkaloids," *Chem. Rev.* **65**, 69 (1965).

S 34 J. E. Saxton, "The indole alkaloids excluding harmine and strychnine," *Quart. Rev.* **10**, 108 (1956).

S 35 E. S. Stern, "Synthetic approaches to the morphine structure," *Quart. Rev.* **5**, 405 (1951).

S 36 M. Shamma and W. A. Slusarchyk, "The aporphine alkaloids," *Chem. Rev.* **64**, 60 (1964).

S 37 K. L. Stuart and M. P. Cava, "The proaporphine alkaloids," *Chem. Rev.* **68**, 321 (1968).

S 38 M. Shamma, *The Isoquinoline Alkaloids–Chemistry and Pharmacology*, Academic Press-Verlag Chemie, New York-Weinheim, 1972.

S 39 A. I. Scott, "Biosynthesis of indole alkaloids" in H 17: p. 105.

S 40 T. M. Sharp, "The alkaloids of *Alstonia* barks. Part II. *A. macrophylla*, Wall., *A. somersetensis*, F. M. Bailey, *A. verticillosa*, F. Muell., *A. villosa*, Blum.," *J. Chem. Soc.* 1227 (1934).

S 41 H. R. Schütte, "Isochinolin-Alkaloide" in M 4: p. 367.

S 42 G. F. Smith and M. A. Wahid, "The isolation of (±)- and (+)-vincadifformine and of (+)-1,2-dehydroaspidospermidine from *Rhazya stricta*," *J. Chem. Soc.* 4002 (1963).

S 43 R. M. Silverstein and G. C. Bassler, *Spectrometric Identification of Organic Compounds*, 2nd ed., Wiley, New York, 1967.

S 44 E. Späth and E. Adler, "Zur Konstitution des Konhydrins," *Monatsh. Chem.* **63**, 127 (1933).

S 45 C. Schöpf, E. Schmidt, and W. Braun, "Zur Kenntnis des Lupinins (Bemerkungen zu der Arbeit von K. Winterfeld und F. W. Holschneider: Über die Konstitution des Lupinins, 1. Mitteil.)," *Ber. dtsch. chem. Ges.* **64**, 683 (1931).

S 46 M. Shamma, J. A. Weiss, S. Pfeifer, and H. Döhnert, "The stereochemistry at C-14 for the rhoeadine-type alkaloids," *Chem. Commun.* 212 (1968).

S 47 R. V. Stevens and M. P. Wentland, "Thermal rearrangement of cyclopropyl imines. IV. Total synthesis of dl-mesembrine," *J. Am. Chem. Soc.* **90**, 5580 (1968).

S 48 F. Schneider, K. Bernauer, A. Guggisberg, P. van den Broek, M. Hesse, and H. Schmid, "Synthese des (±)-Oncinotins," *Helv. Chim. Acta* **57**, 434 (1974).

S 49 M. Shamma, "The spirobenzylisoquinoline alkaloids" in M 1: **13**, 165 (1971).

S 50 P. A. Stadler and P. Stütz, "The ergot alkaloids" in M 1: **15**, 1 (1975).

S 51 J. E. Saxton, "Alkaloids of *Mitragyna* and related genera" in M 1: **14**, 123 (1973).

S 52 J. E. Saxton, "Alkaloids of *Picralima* and *Alstonia* species" in M 1: **14**, 157 (1973).

S 53 V. Snieckus, "The *Securinega* alkaloids" in M 1: **14**, 425 (1973).

S 54 M. Shamma and R. L. Castenson, "The oxoaporphine alkaloids" in M 1: **14**, 226 (1973).

S 55 E. Schlittler and J. Hohl, "Über die Alkaloide aus *Strychnos melinoniana* Baillon," *Helv. Chim. Acta* **35**, 29 (1952).

S 56 C. Szántay, "Structure and synthesis of ipecac alkaloids," in G. Fodor, Ed., *Recent Developments in the Chemistry of Natural Carbon Compounds*, Vol. 2, Hungarian Academy of Sciences, Budapest, p. 63 (1967).

S 57 C. Szántay, L. Szabó, and G. Kalaus, "Stereoselective total synthesis of (+)-vincamine," *Tetrahedron Lett.* 191 (1973).

T 1   G. Trier, *Die Alkaloide*, 2nd ed., Bornträger-Verlag, Berlin, 1931.

T 2   W. I. Taylor, "The ajmaline-sarpagine alkaloids" in M 1: **8**, 789 (1965).

T 3   W. I. Taylor, "The ajmaline-sarpagine alkaloids" in M 1: **11**, 41 (1968).

T 4   R. B. Turner and R. B. Woodward, "The chemistry of the *Cinchona* alkaloids" in M 1: **3**, 1 (1953).

T 5   W. I. Taylor, "The chemistry of the 2,2′-indolylquinuclidine alkaloids" in M 1: **8**, 238 (1965).

T 6   W. I. Taylor, "The chemistry of the 2,2′-indolylquinuclidine alkaloids" in M 1: **11**, 73 (1968).

T 7   W. I. Taylor, "The *Pentaceras* and the eburnamine (*Hunteria*)-vincamine alkaloids" in M 1: **8**, 250 (1965).

T 8   W. I. Taylor, "The eburnamine-vincamine alkaloids" in M 1: **11**, 125 (1968).

T 9   W. I. Taylor, "Indole alkaloids, an introduction to the enamine chemistry of natural products," Pergamon, New York, 1966.

T 10   W. I. Taylor, "The *Vinca* alkaloids" in M 1: **8**, 272 (1965).

T 11   W. I. Taylor, "The *Vinca* alkaloids" in M 1: **11**, 99 (1968).

T 12   M. Tomita, "Die Alkaloide der Menispermaceae-Pflanzen," *Fortschr. Chem. org. Naturst.* **9**, 175 (1952).

T 13   W. I. Taylor, "The *Iboga* and *Voacanga* alkaloids" in M 1: **8**, 203 (1965).

T 14   W. I. Taylor, "The *Iboga* and *Voacanga* alkaloids" in M 1: **11**, 79 (1968).

T 15   S. Tobinaga, "A review: Synthesis of alkaloids by oxidative phenol and nonphenol coupling reactions," *Bioorg. Chem.* **4**, 110 (1975).

T 16   B. Tursch, D. Daloze, and C. Hootele, "The alkaloid of *Propylaea quatuordecimpunctata* L. (Coleoptera, Coccinellidae)," *Chimia* **26**, 74 (1972).

T 17   B. Tursch, D. Daloze, M. Dupont, C. Hootele, M. Kaisin, J. M. Pasteels, and D. Zimmermann, "Coccinelline, the defensive alkaloid of the beetle *Coccinella septempunctata*," *Chimia* **25**, 307 (1971).

T 18   J. Tomko and Z. Votický, "Steroid alkaloids: The *Veratrum* and *Buxus* groups" in M 1: **14**, 1 (1973).

T 19   R. Tschesche and E. U. Kaussmann, "The cyclopeptide alkaloids" in M 1: **15**, 165 (1975).

T 20   W. I. Taylor and A. R. Battersby, "Oxidative coupling of phenols," Dekker, New York, 1967.

T 21   B. Tursch, D. Dalzole, J. C. Braekman, C. Hootele, A. Cravador, D. Losman, and R. Karlsson, "Chemical ecology of arthropods, IX. Structure and absolute configuration of hippodamine and convergine, two novel alkaloids from the American ladybug *Hippodamia convergens* (Coleoptera-Coccinellidae)," *Tetrahedron Lett.* 409 (1974).

T 22   C. Thal, T. Sévenet, H.-P. Husson, and P. Potier, "Synthéses de la (±)-déséthylvincamine et de la (±)-vincamine," *C. R. Acad. Sci.* **275C**, 1295 (1972).

U 1   M. R. Uskoković and G. Grethe, "The *Cinchona* alkaloids" in M 1: **14**, 181 (1973).

V 1   H. J. Veith and M. Hesse, "Thermische Reaktionen im Massenspektrometer. Modellreaktionen zur Untersuchung thermischer Umalkylierungen," *Helv. Chim. Acta*, **52**, 2004 (1969).

W 1   W. C. Wildman, "Amaryllidaceae alkaloids" in P 1: p. 151.

W 2   W. C. Wildman, "Alkaloids of the Amaryllidaceae" in M 1: 6, 289 (1960).

W 3   W. C. Wildman, "Alkaloids of the Amaryllidaceae" in M 1: 11, 307 (1968).

W 4   W. C. Wildman, "Colchicine" in P 1: p. 199.

W 5   W. C. Wildman, "Colchicine and related compounds" in M 1: 6, 247 (1960).

W 6   W. C. Wildman and B. A. Pursey, "Colchicine and related compounds" in M 1: 11, 407 (1968).

W 7   J. T. Wróbel, "*Nuphar* alkaloids" in M 1: 9, 441 (1967).

W 8   F. L. Warren, "*Senecio* alkaloids" in M 1: 12, 246 (1970).

W 9   F. L. Warren, "The pyrrolizidine alkaloids," *Fortschr. Chem. org. Naturst.* 12, 198 (1955).

W 10   F. L. Warren, "The pyrrolizidine alkaloids," *Fortschr. Chem. org. Naturst.* 24, 329 (1966).

W 11   K. Wiesner, "Structure and stereochemistry of the *Lycopodium* alkaloids," *Fortschr. Chem. org. Naturst.* 20, 271 (1962).

W 12   K. Wiesner and Z. Valenta, "Recent progress in the chemistry of the Aconite-*Garrya* alkaloids," *Fortschr. Chem. org. Naturst.* 16, 26 (1958).

W 13   B. Witkop and C. M. Foltz, "Studies on the stereochemistry of ephedrine and ψ-ephedrine," *J. Am. Chem. Soc.* 79, 197 (1957).

W 14   J. C. Willis, *A Dictionary of the Flowering Plants and Ferns*, 17th ed., Cambridge University Press, 1966.

W 15   E. W. Warnhoff, "Peptide alkaloids," *Fortschr. Chem. org. Naturst.* 28, 162 (1970).

W 16   E. Winterfeldt, "Stereoselektive Totalsynthese von Indolalkaloiden," *Fortschr. Chem. org. Naturst.* 31, 469 (1974).

W 17   E. Winterfeldt, "Neuere Aspekte in der Chemie der Indolalkaloide," *Chimia* 25, 394 (1971).

W 18   E. Wenkert and B. Wickberg, "General methods of synthesis of indole alkaloids, IV. A synthesis of dl-eburnamonine," *J. Am. Chem. Soc.* 87, 1580 (1965).

Y 1   K. Yamaguchi, *Spectral Data of Natural Products*, Vol. 1, Elsevier, Amsterdam, 1970.

Y 2   S. Yamamura and Y. Hirata, "The *Daphniphyllum* alkaloids" in M 1: 15, 41 (1975).

Z 1   B. Zsadon and P. Kaposi, "(+)-Vincadifformine from *Amsonia tabernaemontana* Walt. (Apocynaceae)," *Tetrahedron Lett.* 4615 (1970).

Z 2   B. Zsadon, M. Rákli, and R. Hubay, "Indole and indole derivatives, VII Tabersonine from the seeds of *Amsonia tabernaemontana* Walt. (Apocynaceae) grown in Hungary," *Acta Chim. Acad. Sci. Hung.* 67, 71 (1971).

Z 3   K. Ziegler, H. Eberle, and H. Ohlinger, "Über vielgliedrige Ring-systeme. I. Die präparativ ergiebige Synthese der Polymethylenketone mit mehr als 6 Ring-gliedern," *Liebigs Ann. Chem.* 504, 94 (1933).

Index

217